LIVING OFF THE GRID

The Complete Guide for a Sustainable, Tranquility and Simple Life,

a Living of Minimalism and Self Reliance

By

John Edwards

The information herein is offered for informational purposes solely and is universal as so. The presentation of the information is without a contract or any type of guarantee assurance.

The trademarks that are used are without any consent, and the publication of the trademark is without permission or backing by the trademark owner. All trademarks and brands within this book are for clarifying purposes only and are owned by the owners themselves, not affiliated with this document.

TABLE OF CONTENTS

1.

WHAT IS OFF-GRID LIVING?

Around a similar time, every month, a considerable number of people go to their post boxes looking for the manually written letter or their preferred magazine just to be welcomed by white envelopes with small-scale cellophane windows. We are all acquainted with these mailers – electricity, water, gas, and phone charges, all scheming to take your well-deserved cash. For the vast majority, taking care of utility-bills is a tedious and disappointing undertaking. Consider the possibility that there was an approach to become free of these commercial utilities and use your own sources to produce your energy. Overall, there is. Going "off-grid" is turning into an inexorably mainstream decision for individuals hoping to lessen their carbon impression, state their autonomy, and maintain a strategic distance from dependence on non-renewable energy sources.

"The grid" is a common name for the power-grid framework – the connected framework that conveys power to the majority. A typical house is associated with electricity, petroleum gas, water, and phone lines. Going off-grid implies disregarding these open utilities and creating your own energy sources. Some homeowners have decided to be partly off-grid by providing their power and jettisoning their telephone line while depending on the city water and sewage. Others choose to live totally off the grid by digging wells or utilizing a storage framework to gather water. A septic tank deals with the sewage and, much the same as that, no more water bill either.

"Off-the-grid" refers to houses that are self-contained, which means that you don't depend upon natural gas, water supply, electricity, sewers, or other such municipal services. An autonomous home operates independently of all such civil/public services.

We need to dispose of the probable idea that everybody needs to bring home the bacon. Today one in 10,000 can make a mechanical achievement efficient for supporting everything else. Youngsters today are more right than wrong to perceive this foolish idea surrounding earning a living. We keep on making occupations because of the misinterpretation that everybody must be utilized on a type of job because, as indicated by the Malthusian-Darwinian hypothesis, one must legitimize one's entitlement to exist.

As intrigue develops in self-supporting and dreaded reliance on expanding gas imports, an ever-increasing number of individuals want to move "off the grid."

Pioneer organizations are as of now assembling radical, innovative eco-towns, with waitlist records for these houses numbered in the thousands. Innovation is more brilliant, more productive, and less expensive than any other time in recent memory, making these eco-homes a reality for the individuals who can manage the cost of it. Nevertheless, has the innovation become customary and affordable enough for all?

Our gas and power supply frameworks, called systems, are concentrated frameworks that circulate energy where it is required. Market interests are deliberately coordinated, and the system regularly buys an abundance of sustainable power sources from sellers to keep it completely provided during times of respite.

Disconnecting yourself from this system implies losing that wellbeing net – and this has, for quite some time, been the issue for those with desires outside the system. As of not long ago, an off-grid life implied severely constraining energy utilization when the sun was not shining, or there wasn't much wind. Presently, energy-conserving innovation is turning out to be progressed to such an extent that we can store overabundance of energy from the sun to be used at nights, instead of demanding the additional energy from the system. Nevertheless, since the capacity issue is being settled, the subject of whether we can create enough energy remains.

The way to creating enough energy to live off the network is to utilize a scope of arrangements. The energy utilization of the average family changes is determined by their location. On the American landmass, for instance, it is around 30-kilowatt hours out of each day, yet in Hawaii, it is merely 1/2 of that. In colder nations like the UK, where the typical family utilizes 125-kilowatt hours daily, warming homes require a great deal of energy.

In any case, there are numerous choices for keeping us warm, the most effortless of which is consuming biomass (wood and another natural issue). Solar panels and geothermal warming systems permit us to utilize warmth from our environmental surroundings to warm water systems. These can be very costly to purchase, cost a few thousand pounds/dollars, however, are successful and, similar to a long haul venture, will pay off after a while. It is likewise conceivable to change utilized cooking oils into environmental biodiesel items to be used as warming oil or vehicle fuel.

Escaping the gas system can even assist us in taking care of two issues simultaneously. From puppy poop to a renowned London Fatberg, organizations are utilizing our waste – as are you. An anaerobic digester prepared house will change your fecal and water waste into enough gas to cook your dinners. At around £ 700, any reserve funds in the gas created would not earn back the original investment for a considerable length of time, yet it is an extraordinary method to deliver biogas and composts while treating waste.

2.

AUTONOMOUS OR CONNECTED HOUSE: WHAT ARE THE POSSIBILITIES?

More and more individuals dream of autonomy. Beyond the wisdom or not of this aspiration, beyond the level of independence, what are the rights and/or obligations? Are there differences between houses under construction, therefore subject to a building permit application and existing homes, already connected or not, already equipped or not with sanitation systems? Is it possible for houses already connected to terminate these connections? If so, what are the consequences? What are the risks of not following any rules and obligations? Of the many questions, we hope to answer them most objectively and exhaustively possible.

However, there are many exceptional cases, linked to grouped, semi-collective installations, or services offered or not, some cases not falling under the ordinary rules, may arise here or there. We will approach the subjects by themes like water, electricity, sanitation, and will subdivide these themes in different situations, existing houses, or new constructions.

Electricity

Electricity is what seems the simplest in terms of connection or not. There are many possibilities for producing electricity, ranging from solutions that we, of course, do not recommend, in any case continuously or even regularly, from gasoline, gas, or diesel generators to photovoltaics, biomass, wind, hydraulics, etc.

The situation will be different depending on whether it is self-consumption without resale of excess production or production supplemented by a connection to the network. Such installations are framed in terms of power and must be subject, depending on the situation of the property (close or not to a historic monument, close to a site classified as a natural park, etc.) to the prior declaration or based on the building permit.

Construction of houses

The connection of the house to be made, to the network, is compulsory, as per local authorities in many countries. On the other hand, "connected" will not be required to subscribe to a supplier. They will, therefore, have a meter but no consumption. This situation applies to all construction subject to PC (Building Permits), whether it is a new house or an extension subject to PC. Therefore, if work subject to PC is carried out in an existing house not connected to the network, it will be necessary to request the connection to the network.

Existing houses

The occupation of an existing house not connected to a network is possible without connecting it (except the case mentioned above). If a request is made and no electrical network passes nearby, it can be carried out, but the costs of the network extension will be borne by the requester, which can quickly prove to be very expensive.

The resale of an existing house is only possible if the property is connected to the network. This situation does not oblige the ERDF to carry out the network extension. They are and remain the responsibility of the buyer, this notwithstanding the agreement of the resale contract.

Potable water

Water is the most necessary resource for living in a place. Without it, no ability to drink, to simply and adequately wash, to produce food by cultivation or gardening, except to bet on regular and sufficient rainfall. From there, is it possible to live in a place not connected to a supply network? Yes, it is entirely possible, unless local regulations prohibit it.

The municipalities and other entities in charge of water distribution should supply water to all legal constructions, this for the non-private part (to put it simply, either cadastral or duly constructed and intended for housing). In other words, refusal is possible for illegal constructions such as temporary housing. It is also so in the event of a change of the purpose of a building, passing it, for example, from agricultural building to habitable building.

The possibility of refusal does not necessarily lead to the denial; it is therefore strongly advised to make a prior request for a Certificate of Town Planning and/or connection before embarking on a project at risk of refusal.

Construction of houses

It is only possible to build a house, in any case legally, after obtaining a building permit. Whereas this is only granted under the conditions of a possible connection, there will be no difficulty concerning water if a network exists, the PC (subject to compliance with other obligations) will be obtained. If there is no nearby network, the PC will be refused. Connection to the system is compulsory, but water consumption is not, which will not prevent, possibly, the invoicing of a minimum fixed consumption, real or not.

If there is no nearby network and, on request for a waiver, a PC is obtained, it will be possible to dig a well. It will be necessary to do so legally. It is enough to declare the construction of a well or drilling and, of course, to obtain the authorization. If the well does not give water, what can be done? It is possible to use storage tanks, which, in sanitary terms, requires great filtration precautions.

Existing houses

Many existing houses are scattered in cities, towns, and countryside. If they are inhabited or were inhabited after December 30, 2006, they are connected to the public water distribution network. If this is not the case, it is necessary to request a Certificate of Town Planning before any other steps. If it is refused, for example, because it is too far from any network, we advise you to approach the town hall and start a negotiation, without prejudging its outcome.

If this house has a well, it will be possible to live there (provided that its water is drinkable); otherwise, you may need to prepare to abandon any housing project of this house, except, as described above and on condition that it is possible to set it up, live on a storage tank, as long as water is supplied!

Sanitation

Home sanitation means wastewater discharge. There are two categories of discharges: gray water and black water.

Gray water: formerly the water discharged by washing dishes and toilet water (not from toilets!). In both cases, polluted with detergents, in minimal quantities; today: bathroom, kitchen, washing machine, dishwasher are increasing in this type of water discharge.

Blackwater: formerly known as sewage; today, the water released from the toilets.

The treatment of these releases is essential. Indeed, if the sanitation is not correct, these different waters can seriously affect the environment and, above all, cause serious health effects. Once connected to the network, residents who wish can limit their discharges, including those related to toilets, by installing dry toilets at home.

Construction of houses

If a sanitation network exists, it is compulsory to connect to it. Otherwise, it will be necessary to carry out a non-Collective Sanitation installation. In this case, the Public Non-Collective Sanitation Service (PNSS) must be consulted in advance.

Existing houses

These houses can present very different situations, in particular about the Collective Sanitation Network (CSN) or Non-Collective Sanitation (NCS)

Already connected to a CSN

If the house is already connected to the collective sanitation network, the only possibility for the occupants to limit their discharges to this CSN consists of the installation of dry toilets.

House not connected to a CSN

A house not connected to a CSN but which its owner wishes to live in must have a wastewater treatment system.

CSN available

If a CSN is easily accessible, it will be mandatory to connect to it. If, although individual sanitation pre-exists, a CSN is set up, it will be compulsory to connect to it. This obligation will necessarily generate some work; those on will be the responsibility of the owner. He will also have to pay the related taxes. However, it seems that this can be disputed if this connection will not allow its beneficiaries to save money.

CSN not available

Wastewater, both gray water and black water, will need to be treated. The most commonly used system consists of a so-called all-water septic tank (it treats both gray water and black water). Other systems can be envisaged, identical to what has been discussed above, PNSSs. We speak in the plural because they are diverse. Some treat only gray water, and others treat all water. In the case of non-treatment of wastewater, you can opt for dry toilets. In all cases, these solutions must have the approval of local authorities and require monitoring after commissioning.

Rainwater

The rain has always fallen where and when it wanted. It is more and more true and… irregular!

Sanitation-water-fluvial

It becomes more and more necessary to take all measures for its proper management. This is why we advise to drastically limit any sealing surface and to favor systems allowing the penetration of water into the ground.

Construction of houses

Any construction must be provided with means of evacuating the rainwater it collects, at least equivalent to what the situation was before its creation.

It is becoming more and more common, the evolution of town planning obliges, to collect rainwater, in particular, collected by roofs, and organize the retention in collective retention basins. This makes it possible to delay their transit, arriving too quickly and too much in streams and rivers.

It is up to the owner to organize the evacuation of its rainwater at his expense. It is also up to him to organize it in such a way that it is in no way the cause of any flooding, whether in the public domain or neighboring land(s), under normal conditions of rainfall.

Existing houses

Any pre-existing situation regarding rainwater, it must continue, both for the benefit of the owner and about his obligations about the public domain and neighboring land.

Telephone and/or internet networks

Since the advent of cell phones, the connection to communication networks is more and more often motivated not for calling but the use of the web. However, in certain regions, called "white zone " because they are not covered by the cellular telephone network, the wired network is the only possible solution.

Construction of houses

Obtaining a PC is subject to a possible connection to the telecommunications network. It is mandatory to provide for the possibility of connection. However, there is no obligation to request it; in this case, the connection ducts will not be used and will remain available…just in case.

Existing houses

There is no obligation to provide for connection or to connect to the wired telecommunications network. However, with the advent of various services available via this network, we strongly advise, before choosing the location, to find out about the type of cables in place and the possible speeds.

Access

Unless you want to live as a hermit, a house must be accessible. Formerly a pedestrian or mule track could suffice, today the needs are different if only in cases of emergencies.

Home accessibility

The public access route must allow access to these emergency vehicles and resist the successive passage of vehicles. It is the responsibility of the municipalities to ensure their maintenance.

On the other hand, the use is up to the homeowner, while still respecting the limits which can be imposed there, in particular as regards the type or weight of authorized vehicles.

Construction of houses

In the absence of possible access for emergency services via a public road to the entrance to the site, the CP will not be granted. The beneficiary is responsible for carrying out the private access portion.

There are no specifications regarding its dimension, nor the type of surface. However, let us not forget that overuse and depending on the climatic conditions, it must withstand the numerous passages of vehicles but also allow rainwater management. In order not to generate new water collection and thus further increase the risk of flooding, it would be good if it could continue to infiltrate into the ground, in a swamp, for example.

Existing houses

As for the access of this type of house to the nearest public road, it exists at least by a pedestrian path. This is by choice. Nothing can be demanded from the occupants when the area develops. As the building was purchased, such, at a minimum, it must remain accessible.

Conclusion

In terms of connections and/or access, an owner has rights and obligations. It is up to him to ask for its enjoyment as well as to assume its constraints.

If the house is still at the project or design stage, that is to say, planning, whatever may be the hopes that the future owner places there, we can only advise him to obtain all the necessary information now rather than risking setbacks later on.

If the dream is at the level of an existing house to buy, i.e., if it is not yet fully owned, we advise the future owner just as strongly to do the same. The minimum, in case of doubt, is to request a planning certificate.

3.

HOW TO CHOOSE THE RIGHT PROPERTY

Types Of Ecological Houses

Now is the time for prevention and changes in habits towards our sweet planet. Whether it is in our daily life or thanks to associations, every gesture counts. But we sometimes forget that our house can also be better adapted to its environment. So, we decided to teach you more about the different types of greenhouses! You can take inspiration from it for your future constructions and, thus, please Mother Nature!

Low Energy Buildings (Bbc)

Let's start with the BBC label first! Created in 2007, it asks households, wanting to obtain it, to minimize their energy usage for reducing the environmental pollution. This label requires, with new buildings, an energy standard of 50kWh per square meter per year. Regarding existing buildings, they must be less than 50% compared to their current consumption. To benefit from this label, we take into account it's heating, air conditioning, lighting, domestic hot water, and ventilation. A small advantage, you will benefit from a lower property tax and will please your environment, so you might as well enjoy it! However, other more demanding standards have been put in place to take care of our planet. If you want to continue reading, we will reveal the secrets of a greener home!

Bio-climatic house

You cannot define a bioclimatic home with a specific example. The goal of this type of ecological house is to adapt to its environment, its advantages and disadvantages! Its only source of energy is based on outdoor climatic conditions; it's time to separate yourself from your heating and air conditioning systems! It is imperative to take into account the environment, materials, and orientation of the house before starting your work. You must resist the cold of winter as well as the heat of summer without using technologies harmful to the planet. Forget the windows and the living rooms on the north side. On the south side of your home; large insulating windows are essential to benefit from the heat of the sun. Above all, use ecological materials! Hemp or cellulose wadding for your insulation and wood or terracotta for the finishes are very popular. It's up to you to find the right tips to benefit from thermal comfort following this principle of bioclimatic house. Now you just have to get started!

The passive house

Three criteria are necessary for your home to be considered passive. Initially, your final energy requirement should not exceed

50kWh per square meter per year. Then your airtightness must be at n50 <0.6 h-1. Finally, your heating need must be less than 15kWh per square meter per year. Besides, you will have to respect certain principles, such as strengthening the insulation by removing thermal bridges and limiting household appliances. Without forgetting to install double-flow ventilation, which allows the renewal of the air in a more natural way! Choose, as for the bioclimatic house, windows on the south side to capture the heat and maintain it in your home. Everything you produce, you should recycle! For this kind of habitat, it is preferable to opt for a simplistic architecture to avoid as much as possible a so-called "energy-consuming" effect. But after all, why make things complicated when you can make it simple and better?

The positive house

The keyword for this kind of ecological housing? Autonomy! The positive houses further and healthier. You will produce your own energy (if not more) to satisfy the whole house and its inhabitants. So, think of bringing technologies adapted to renewable energies, just like the passive house. Solar panels, heat pumps, rainwater collectors, and so on, are at your entire disposal to facilitate the ecological transition of your home. However, it should not be considered as being self-sufficient 24 hours a day. It is not without risk that your solar panels will no longer produce energy due to a downpour! But do not panic, once the sun returns, it is possible to store the surplus of unused energy. As far as materials are concerned, you are not restricted, as long as they are ecological and natural. But when it comes to architecture, don't think too big and too complicated! You must adapt to your environment, the climatic advantages and disadvantages, and the type of terrain.

After these few explanations, you just have to build your type of greenhouse! Innovation is sometimes a source of inspiration; do not hesitate to find out about new construction materials and tools that favor the environment. Finally, break the cliché, which emits the idea that an ecological house is not accessible for all and gets started!

Please note: you will need planning permission for the construction of your house. Also, to make your life project happen as quickly and efficiently as possible, we offer an online building permit service! Describe your project, an urban planning expert, and a facilitator who will take care of your building permit request!

Earthship: the autonomous house?

An Earthship is a house built with 40% recycled products, namely used tires, packed down and buried so that they generate no emissions, but also boxes or glass bottles, for example.

In this house concept, no drop of water is spoiled as explained by one of the owners of Earthship house, Benjamin, since the rainwater is collected, it allows you to take showers (heated by solar panels and sometimes the water is too hot d 'elsewhere), to operate all the devices (dishwasher, taps of all kinds...) then this gray water feeds the pool which is filtered by the plants in the green/autonomous house which allows it to be reused to create black water (toilets…) Which is then sent to a treatment plant.

In the same way, the house has no heating and works in geothermal energy and ensures a temperature of 18 to 23 degrees all year round whatever the external conditions if it is inhabited and that specific basic rules are respected.

Before taking action, Benjamin had the opportunity to live in an Earthship for a few days in California. This allowed him to realize that it was entirely possible for him and his family. But, in reality, he was already engaged in a personal ecological transition for many years. This decision to build this house did not come at a whim, and it is in this same logic (brought as emphasized above by financial considerations) to leave the urban world and join the rural world. Benjamin explains that he no longer wanted to have "his ass between two chairs" and decided to be consistent with his philosophy. However, that does not prevent him from traveling, from going back to the city when necessary, of course.

We could criticize him for not being totalitarian since taking the train is spending electricity.

But modifying one's habitat and, therefore, the place where we spend most of our time and ensuring that we produce as much of our food as possible is already a huge step.

Adapt according to nature and not vice versa

They are forced to do according to nature, that is to say, according to sunshine, rain or even heat and cold, but it is a symbiosis with nature, which is very important to Benjamin and his family. Their comfort, therefore, depends on nature, and they do not try to force it for their comfort. For me, they are proof that you can change your lifestyle and have a different ecological impact while not denying the essentials. The real question ultimately, as Benjamin points out, is whether you need to be in a physical environment that does not allow you to have daily mobility.

But it is true that for many when we see this kind of project, it makes us think that it is great but not for us and we are probably satisfied with small gestures of everyday life. Benjamin's message, of course, is to say and prove that this is completely possible without necessarily being a hippie or a revolutionary. Building such a house involves particular choices such as sunshine, which will determine the inclination of the wall of the green/autonomous house to ensure the optimization of the green/autonomous house yield. The Earthship is an ecological solution and an alternative to traditional houses. Fully autonomous in water and electricity, the Earthship is still too little known in France. However, the advantages of this habitat are manifold. Here are five reasons to live in an Earthship:

1) Living in Earthship means choosing to respect the environment and living independently

Earthships are made from ecological and recycled materials. The impact on the environment is, therefore, minimal. Earthships are passive houses: energy is supplied by the sun using solar panels; rainwater collected, filtered, and treated directly on site. The presence of a green/autonomous house makes it possible to grow fruits and vegetables in any season. It is also possible to raise chickens there for eggs. The interior temperature remains constant in summer and winter because the Earthship is partially buried, and the Earth acts as a natural heat regulator.

2) Living in an ecological house does not mean living without comfort

When people hear about an ecological house, they imagine houses without modern comfort. However, living in Earthship does not mean living in a cold and humid cave. On the contrary, Earthships have everything you need to live well. The rainwater collector provides access to running water. Solar panels provide electricity. Above all, the Earthship is custom built to meet your needs and your desires.

3) Materials To Use

The materials are simple to find. The main wall of an Earthship is made up of used car tires filled with soil. The non-load-bearing interior walls are made of aluminum cans and glass or plastic bottles.

Construction is quick and requires no special skills other than willpower and common sense.

4) Cheap

The materials used are, for the most part, recycled materials. These materials are very easy to find. We can, for example, collect used tires for free and in large quantities in Auto Centers, or classified ads. The earth to make the walls is taken directly on site.

5) Living in Earthship means choosing originality

Finally, living in Earthship means living in an inexpensive and environmentally friendly house. Fully autonomous, with this alternative accommodation, you will no longer pay any electricity or water bills. Besides, the Earthships are original constructions that you do not come across on every street corner. You will not have the same house as your neighbor, chosen from catalogs. Your Earthship will be entirely in your image down to the smallest corner.

4.

WATER BASICS IMPORTANCE

In this part, we have decided to detail each step be carried out in the necessary order, to collect and use rainwater, daily, at home.

The efficient and automatic installation of the storage and exploitation of rainwater is relatively complex. This is why, before any decision, ask yourself the right questions and gather the correct information. The first questions to consider at the start of such a construction project to choose, size of the equipment, and plan for water recovery are:

- How many people live in the home? Think about the future, the number of children, for example.

- Gather all the information useful for the correct sizing (water bill in particular).

- What roof area do you have? (or the future built surface in case of new construction).

- Estimate the average volume of recoverable water for a year, depending on the city.

- Estimate your needs and then deduce what autonomy will be predestined: partial or total.

- Think about the installation of the tank: inside/outside, underground/semi-underground/surface, location, etc.

- Will there be several gutters to collect with? Where will they come from?

- Locate the places where the water collection pipes (gutters, etc.), pumping, electrical supplies, the arrival of the emergency city water valve (if necessary) will be conveyed.

- Choose the equipment according to the need, do not hesitate to get several quotes from various known and recognized professionals.

The following characteristics are the different answers to the questions to target the example.

The family currently consists of two people. This family will soon be considering two more people. Currently, the volume of water consumed is around 88 m³ of city water per year, considering a daily volume of 0.12 m³ per day and person. Following the arrival of two additional people in the future, they will consume an additional amount of water for their daily use so the volume of water will increase to 175 m³ of city water, per year

Calculation: $0.12 \times 2 \times 365 + 0.12 \times 2 \times 365 = 175.2$ m³

The house will be of the "new house" type, that is to say, that the project to collect rainwater will be created at the same time as the construction of the house. The floor area of the house will be equivalent to 100 m². Also, this house will have a garden of 50 m², of which a maximum of 30 m² (20 m² of concrete/tiled surface) may be available for the installation of the storage tank.

Estimated volume of recoverable rainwater

The volume of water necessary for this family of two people, over a year, is 88 m³. However, this volume can vary from year to year, and that is why we are increasing it by 15%, the new capacity being 101 m³. By consulting the meteorological data for local precipitation in the city of Douai, on average, we can hope to recover 600 mm of rainwater per m², which represents 0.6 m³ of water collected per square meter per year.

If we consider a roof area equal to the floor area of the house (100 m²), we can hope to collect 60 m³ of rainwater per year (precipitation = 0.6 m³/m², calculation = 0, 6 x 100 = 60 m3). We can, therefore, see that the volume of water collected is much lower than the amount of water required for this family of two people.

A choice is essential: increase the roof area to be independent or partially use city water during periods of drought? The decision was to increase the roof area since two more people will arrive in the future. The roof will be enlarged on the sides of the house to create a shelter for firewood and shelter for two "carport" type structures. The new roof surface will be 170 m² in a glazed tile (to collect the maximum amount of water).

Calculation: precipitation = 0.6 m³/m², 0.6 x 170 = 102 m³, so they will have the possibility of collecting 102 m³ of rainwater per year with this surface.

With their current consumption, the use will be of a total type for the family of two people, while this use of rainwater will become partial when the other two people live in this house. Overall, the volume of water consumed will be greater, so the city water network will be used.

Sizing and choice of tank

This dimensioning is related to the consumption of the family and the percentage of rainwater recovery, which is a function of the roofing material.

The tank capacity (C) is based on the volume of recoverable rainwater (V) per year, the annual rainwater requirements (B) and the choice of the number of reserve days (N), that is:

$$C = (V + B)/ 2 \times (N/365)$$

We are targeting an autonomy of 60 days with a full tank to make up for two months without precipitation, i.e., N = 60.

Remember: V = 102 m³ and B = 88 m³

Calculation of the tank capacity

$(170 + 88)/ 2 \times (60/365) = 129 \times 0.16 =$ about 20 m³

We will, therefore, need a tank with a capacity at least equal to 20 m³. But in general, the basic rule is to provide a tank rather too large than too small!

The needs of the family being as well for the exterior use as for the interior use (toilet, kitchen…), it is decided to take a concrete tank, because of the neutralization of the pH by the cement in the tank. If the user had been only towards the outside (garden), the choice of the material of the tank could very well have been plastic because this water would not have needed to have a neutral pH, necessary in the case of the toilet.

There are two types of concrete tanks. The prefabricated container purchased and installed directly in elements or the tank, which is created after having carried out the excavation.

After a study of different quotes to set up a prefabricated tank, we realize that with the costs of installation, elements, and purchase of the container, the price may increase by approximately $ 1,000 compared to the creation of the tank by a masonry company which would take charge of the construction of the tank on the site.

It was chosen to build a concrete tank poured directly into the excavation by our means (formwork + purchase of materials) of 20 m^3 for storage (see previous calculation) and 4 m^3 for the settling part (20% of the volume of the largest). This tank will cost us around $ 2,000 in materials if we build it ourselves (concrete: $16m^3$ or around $ 1,500 + irons + formwork + water repellent). The dimensions of this tank can be 3.5m x 4 m x 1.5m deep for the storage tank and 2m x 1.5m x 1.3m deep for decantation.

The overflow (disconnector siphon) is to be placed below the maximum level of water in the tank, i.e., 20 to 30 cm below the upper slab. It will be necessary to equip it with a non-return valve to avoid backwashing of water, rodents, or other undesirable things in the tank.

Layout and installation of the tank

The installation of a tank is easier in the context of new construction than in that of a renovation. It is more demanding in terms of accessibility concerning the presence or absence of trees or buried networks (e.g., water, gas, electricity). Also, true when the drainage networks have been buried for many years and must be discovered for the connection of the tank. This operation is to be carried out with delicacy. Installation is, therefore, more expensive because it requires more time. When the building is already constructed, the choice of the tank must take into account the different dimensions between the downspouts of gutters and those of the rainwater evacuation network.

The location of the tank must be chosen with the highest care and will be located as close as possible to the technical room (cellar, garage, outbuilding, etc.), which will house the pumping unit. This proximity will limit the pressure losses induced by the section, the length of the pipe, and the difference in level between the suction strainer and the pumping unit. It is recommended to check the characteristics of the soil 2 or 3 meters deep; the presence of rocky soil can make it difficult to excavate the earth from the tank.

In our case, the installation of the tank will be done before the passage of water, gas, and electricity. The soil excavated for the container will be deposited at the bottom of the garden during the construction of the house and removed by a truck later. The dimensions of the excavation must allow installation without touching the walls. The excavation bottom will include a 10 cm thick bed of sand so that the tank has a good foundation over time. A space of at least 30 cm between the tank and the walls of the excavation is required to handle the formwork for concrete. Once the excavation has been carried out, meticulously prepare the formwork and the irons of the tank. This work is vital. It must be done with the greatest care. Then pour the concrete to the bottom of the walls.

Choice of the type of gutter pre-filter

The element that is placed at the exit of the gutters is essential. This element, called a pre-filter, is used to clean rainwater beforehand to rid it of dead leaves, dead birds, etc. so as not to cause heavy unwanted objects in the tank. This pre-filter can be automatically or periodically cleaned; the main thing is that it does not get clogged too often.

Self-cleaning models are more and more common because they are simple and efficient, but expensive (from $300 to 500). This cost can drop to around 50 euros for a standard type pre-filter, with a metal basket, to be cleaned regularly.

The roof area (170 m²) will require a minimum of four downspouts and two manifolds, which will be directed to the location of the pre-filter using a "single tee."

The equipment chosen to fulfill this filtration function will have a cost of around $400, as it will be of the automatic type.

Choice of the suction group

So far, we have only seen "how" to collect rainwater. But how do you get it to circulate in house pipes under sufficient pressure? Only with a suction group, which is the most essential equipment, after the tank. More generally, in France, this system is called a self-priming suppressor group, because it draws water from the tank to send it back into the house network at a service pressure of 3 bars.

Some different types of equipment allow us to fulfill this function; we will only deal with the self-priming booster, which is the most widespread equipment for a very satisfactory quality/price ratio. The other pumping systems are:

- Piston pumps.

- Centrifugal pumps.

- Self-priming side channel pumps.

- Pumps for semi-deep wells.

- Pumps with hydro-ejector.

- Submersible electric pump groups.

The system chooses a tank; it is autonomous and does not start until after the use of some 10 liters and a drop-in pressure on the network. This allows the pump not to be always stressed. This type of product has a cost of $ 100 to $ 1,000, depending on the material of the tank and its capacity. The larger the capacity of the container, the less demand will be placed on the pump because a reserve of pressure will be stored therein. The flow rate of this type of pump is approximately 4,000 liters per hour.

For the installation, we will choose a booster with characteristics similar to these:

Pressure on/off triggering of the pressure switch: minimum 2.1 bars, maximum 3.5 bars.

The 20 to 60-liter tanks are intended for relatively occasional use (a family of 1 to 6 people), while the 100 to 300-liter tanks are intended for regular use (family of more than six people, hotel, etc.). To ensure a significant pressure reserve in the event of simultaneous tripping of several taps and to prevent the pump from starting too often, a 60-liter tank will be chosen.

The power of the pump will be 1450 W; this power varies in proportion to the capacity of the tank.

The flow will be around 3,900 liters/hour with a maximum delivery head of 20 meters, depending on the characteristics of the pump chosen.

The control panel

The control panel is an automated central control unit using a microprocessor. The electronics control and manage the entire installation. It allows the automatic supply of water to the city network when the rainwater runs out.

The control panel consists of a reading of the water level in the tank, a meter, a solenoid valve, a backflow preventer, a distribution manifold, as well as a simple priming pump regular. Each of these elements can be chosen independently, to be assembled on the desired installation. In our case, it was decided to directly install a suppressor unit equipped with a 60-liter pressure reserve tank, instead of the simple pump that starts each time water is used at a tap.

The control panel has an estimated cost of $2,000 without a pump. This control panel will be used to automatically control the installation, regardless of the level of rainwater in the tank. The changeover between the city water network and the container will be done automatically.

Principle of operation

A Three-way electromechanical valve controlled automatically depending on the resource (possibility of manual control).

(1) Booster for rainwater supply, at a pressure similar to the drinking water network (for installation, this will be replaced by the 1450w - 60-liter booster group).

(2) Tank level probe.

(3) Command and control panel.

(4) City water buffer reserve supplying the booster pump in the event of a lack of water in the tank (reserve regularly renewed to guarantee its quality).

This type of equipment is very expensive, at around $ 2,000. However, it perfectly meets the standards of disconnection imposed by law.

This control panel will be useful because of the replacement of the lack of rainwater directly by city water.

Choice of filters

After choosing the type of containment, you must also consider the appropriate type of filtration. It is a question of detailing the different types of filters that exist and their function. In this installation, we talk about the choices made for this house, why, as well as the cost.

Rainwater falls on the roof, flows to the gutters, then passes through a first pre-filter, which begins to filter the water roughly. It, therefore, prevents leaves, dead birds, etc. from passing when the water goes to the storage tank.

To ensure water quality, it is necessary to filter it. It is possible to combine different filters depending on what is expected. After having carried out several quotes according to several choices of filters which were envisaged, it was chosen in this construction to set up an advanced filtration system which certifies an excellent filtration of the water, which will go as far as human consumption (food).

When water is needed, the water is drawn into the tank and then sent to a 40 μm filter to obtain finer filtration and to protect the booster group from large particles sucked in by the strainer. The cost of this filter is $100. Then the water is directed to the booster. After this equipment, it is decided to put a cartridge filter and, more particularly, a 25 μm particle filter. The installation of such a filter, at the outlet of the booster, is used to filter the particles as water is consumed and at 25 μm to allow filtering progressively. The filtration is mechanical, and it is only filtered during use. Behind this filter, install an activated carbon filter is a good supplement because it eliminates organic matter and any bad odors. The cost of these two filters is $250. These two filters allow the filtered water to be used for needs of which an "exceptional" quality is not required. We can, therefore, supply the toilets and the garden with this type of filtration, but be aware that we could also supply the shower, the washing machine, and the dishwasher. To be sure that the water is of good quality, it is decided to invest in a more efficient system at the cost of around $800 which sterilizes the water for use in the kitchen and the bathroom; this system is a germicidal, bactericidal filter which allows the disinfection of water by a UV lamp.

Choice of other devices

Insofar as the main equipment has been mentioned above, the other equipment required is:

- Strainer fitted with a float (to draw water from the tank to the booster group);

- Flexible hose for strainer;

- Aquarium aerator (to oxygenate the water in the tank);

- Other pipes for house connections to the distribution network;

- Anti-turbulence tube;

- Wire rack;

- Tank aerator (external evaporation pipe);

- Cast iron manhole (allows the passage of heavy vehicles);

All of this sub-equipment represents around $ 1,500. Depending on the quality of the aerator, strainer, etc. this price can be doubled.

Establishment of plans

To better understand the location of the various elements and equipment making up the installation, several plans are necessary. Plans can be drawn by hand. You don't have to draw plans with software that costs a fortune if you do the installation yourself.

Out of all the water consumed by this family per day, only a few liters of food quality are needed. Of all the water consumed, 20% is used for flushing the toilet, and 12% is used for the garden and washing the floors.

For non-food uses, we especially need freshwater, that is to say, little limestone and little loaded with mineral salts. It must be harmless in case of accidental absorption of small amounts. Hence, this other question: should all of the water be made drinkable, at great cost, to use 60% for food?

A solution to avoid drinking of quantities of water that have no use (WC, garden, etc.), we separate the uses. Thus, the garden / WC use can be used directly at the outlet of the suppressor filter (32%); the other uses (68%) will be made drinkable. As a result, the cost of drinking water, maintenance, and use will be lower.

Rainwater treatment

Depending on the type of equipment chosen, and its quality level, the total cost of installation can vary considerably.

5.

INDIVIDUAL (AUTONOMOUS) SANITATION: HOW DOES IT WORK?

E very day, you use water for dishes, showers, laundry, toilets that produce "domestic wastewater." Polluted, they must be purified before being discharged into the natural environment. This is the role of the septic tank and the all-water tank.

As a reminder, if your home cannot be connected to the sewer because the sewer does not exist on the street or because your house is isolated, the installation of a septic tank or all-water tank is essential; the first receives only sewage (toilets) while the second also receives gray water (sinks). However, if your house is served by a sewer system, you are required to connect to it. The principle remains the same: wastewater must be treated before reaching the natural environment (soil, stream, ditch)

Well designed and well installed, non-collective sanitation, the cost of which is several thousand euros (including installation), guaranteed for one year (in the event of a leak, for example), respects the environment and public health. Its lifespan is around 20 years.

Although the functioning and the lifespan of a non-collective sanitation system depends on the care taken in its construction and maintenance. This system, for the simplest solutions, includes an all-water tank, that is to say, a concrete or plastic tank that collects water from toilets, sinks, sinks, showers, and bathtubs, as well as a spreading, posed following the pit all waters. It consists of a network of plastic pipes, drilled at the bottom, and buried at shallow depth. It ensures the dispersion of wastewater.

The treatment is provided by bacteria contained in the surface soil. Integrating easily into your land and offering a comfort identical to that of collective sanitation, the pit, septic, or all waters, guarantees a good elimination of pollution at an acceptable cost, but also an effective purification technique that contributes protect watercourses and groundwater. However, the installation must be well designed and correctly performed for effective treatment.

The stages of non-collective sanitation are collection, pretreatment, and processing. As for the first step, you must first collect the wastewater produced in the house (toilet water, kitchen water, bathroom water, washing machine water) to direct it to the pit septic. (Rainwater: rainwater from roofs and terrace water, for example, is not considered as wastewater; it is discharged separately.)

As part of the pretreatment stage, you should know that the water collected waste contains solid particles and greases which must be eliminated in order not to disturb the subsequent treatment, this is why the passage in the septic tank or all water tank is essential. At the exit of the pit, the waters are free of unwanted particles and can, therefore, be treated by the soil.

Finally, concerning the treatment, a device of pipes then of drains makes it possible to bring the pretreated water into contact with the bacteria naturally present in the soil or a mass of sand. It is the action of these micro-organisms which ensure the cleaning of wastewater before dispersion by flow in the subsoil.

The size of the pit varies according to the size of the house and the number of occupants. Thus, for a dwelling of five rooms (rooms = number of bedrooms + 2) or less, the pit must offer a capacity of 3 m^3; 4 m^3 and 5m^3 when the house will include six and seven rooms respectively. So, if you plan to enlarge your house, your pet, and your treatment area, depending on the number of main rooms, will have to change. Be aware in any event that the capacity of a septic tank can exceed 55m^3 (55,000L).

For the device to operate durably, your installation must take into account the characteristics and constraints of your land: characteristics of your soil: permeability, thickness, the possibility of rejection of treated water, etc.; the presence of water: level of the groundwater table (groundwater), the slope of the land, available surface; congestion of the plot (property line, presence of a vegetable patch, access to a garage, etc.), the existence of a nearby well, etc. For example, very impermeable clay soils should preferably receive a bed of sand to allow the evacuation of wastewater.

"Porous soils are well suited for filtration," says Noël Cortinovis of Cortinovis TP, a public works company specializing in sanitation. In fact, before contacting the installer, you can call on an architect or a design office to study the characteristics of the soil and define the appropriate sector, or even contact the Urban Community (SPANC) to obtain information and advice on the procedure to follow and ensuring the various controls. You can also contact notaries in the context of a plot division or sale of a property or at the town hall in charge of issuing town planning requests and respecting safety and public health.

To simplify the procedures, it is therefore strongly recommended to call on a public works company specializing in sanitation for the creation of your autonomous system. Why a public works company? "Because this site involves earthworks involving the use of shovels and mini-excavators, in particular," says Mr. Perez, an engineer at Simon. "Installation is to be avoided in the event of rain. There is no particularly convenient time of year; that said, the most favorable seasons, in theory, are spring and autumn," he says. Note that you are free to buy the equipment yourself. "These works can last from three to eight days depending on the size of the site," says Noël Cortinovis.

As for maintenance, it includes various operations. Indeed, it is advisable to check, every six months, the non-clogging of the grease trap. And every four years, it is advisable to remove the filter material from the pre-filter (in the pit or after the pit), wash it and have the pit emptied by an approved company, a specialized public works company in sanitation.

A small fraction of the sludge, as well as the interstitial water (laden with bacteria), must be left in place after emptying for a rapid restart of the bacteria. Also, note that the public non-collective sanitation service is empowered to ensure the verification of proper maintenance of the installation.

As a result, to benefit from a septic tank or all water for as long as possible, it is recommended to avoid throwing away certain substances (corrosive liquids, oils (fried food, oil changes), paints, medicines, objects likely to clog the pipes) and not to waterproof the surface, avoid storing heavy loads or passing vehicles in it, and install it more than 3m from the trees, as close as possible, however, to the outlet of the cooking water to limit the risk of clogging of the pipe leading from the house to the pit (respecting a minimum distance of approximately 3 meters). Equipment requires good ventilation because the organic matter contained in the septic tank releases by rotting gases such as CH_4 (methane) and H_2S. A poorly ventilated pit generates corrosive gases (sulfur odors), which affect the longevity of the materials.

After impoundment, the bacteria are brought naturally by the feces. These bacteria are present in sufficient quantity to ensure the partial degradation of feces: it would be superfluous to add special activating products. However, in the case of a second home occupied only once or twice a year (for holidays, for example), the use of special activators such as Eparcil, for example, can help restart the pit. Besides, interruptions to the supply of the pit for short periods (holidays), or the exceptional supply of detergents (in reasonable quantity: 1 liter of bleach from time to time) have no major impact on its operation.

Finally, from a regulatory point of view, any owner of a building or individual house not served by the public collective sanitation network must, before installing a non-collective sanitation device, submit a so-called technical file. "request for authorization to install a septic tank" at the town hall which will send it to the Inter-communal Association for Non-collective Sanitation. This technical file must be submitted simultaneously with the request for a building permit or the declaration of works for the installations subjected to one or the other of these procedures, at least one month before the realization of the works for the projects not following specific town planning procedures. Sanitation can only start after the explicit agreement of the town hall.

5 (a). What are the different non-collective sanitation systems?

The different uses of water in a home are a source of pollution, which must be treated to preserve the environment and avoid health risks. Consequently, all of the wastewater discharges must be connected to the non-collective sanitation installation: WC, shower, sink, household appliances, and even the possible sink located in the garage, which is not often used!

This is why we often speak of "all water tank" or "all water septic tank."

The essential steps for the treatment of wastewater

There are a wide variety of installations, but to properly treat wastewater, we must always perform three steps:

1- Collect the wastewater and transport it out of the house, through a set of pipes that lead to the treatment works. It is advisable to make these pipes accessible by manholes, in particular, to allow them to unclog easily in the event of an obstruction.

2- Treat wastewater, using one or more structures (pits, filters, spreading, etc.) that differ depending on the type of sector used. The treatment is done in two stages: one speaks of primary treatment (or pretreatment within the meaning of the regulations) and secondary treatment (or treatment within the meaning of the regulations).

3- Drain the treated wastewater as a priority by infiltration into the ground (A).

If infiltration is impossible and subject to authorization, the evacuation can also be done on the surface (B) towards a ditch, a watercourse, or other. Finally, for the treatment of wastewater to be successful, as in collective sanitation, the "residues" (sludge, waste materials) must be removed and treated specifically. Today, we still find homes that discharge their wastewater into a lost well, more commonly called a sump, preceded by a pit or not. However, this type of arrangement does not allow wastewater to be properly treated: it is only a means of discharging it. This is why lost pits have been prohibited by regulation for many years.

5(b). Ask yourself the right questions before choosing the installation!

The Ministry of Ecological and Solidarity Transition has published a guide to support individuals in the choice of their non-collective sanitation installation with several questions to ask: number of rooms, available area, and constraints of the land that will host the installation, maintenance constraints, noise pollution, financing.

The soil and industry study must allow the optimal solution to be chosen from a regulatory and "technical-economic" point of view, depending on the constraints of the plot and the building. For any request for the installation of a sanitation device, this study is one of the compulsory documents.

What are the processes that treat wastewater?

As in collective sanitation, the treatment of wastewater is ensured by bacteria that digest organic matter in the presence of oxygen, which generates sludge, gases, and treated water. These are biological processes that work naturally and are maintained as soon as the bacteria are "nourished" by the supply of wastewater.

There are two phases of treatment:

1. **Primary treatment**

This is a first purification phase which combines:

- Physical processes: retention of floating materials such as grease and settling of the solid materials which form the "sludge"

- Biological processes: digestion of accumulated sludge

2 Secondary treatment

It is provided by purifying bacteria, which will finalize the purification of wastewater.

If the purification processes are generally the same, the means of implementing them differ depending on the sector. The following illustrations – extracted from the information guide on facilities for users, drawn up by the ministries in charge of ecology and health – show the functioning of the different families of supply chains.

The case of traditional supply chains

In the case of traditional systems, the primary treatment is provided by an all-water tank equipped with a prefilter (integrated or not in the tank), which makes it possible to trap the solid matter not retained by the tank. In some cases, the tank is completed with a degreasing tank.

There are two families of traditional industries:

Pits and spreading in the soil in place, for which the secondary treatment is ensured by the bacteria present in the natural ground

(one speaks about "trenches of spreading")

Pits and spreading in reconstituted soil: in this case, secondary treatment is provided by bacteria present in soil reconstituted by sand (case of the sand filter, drained or not).

The case of approved supply chains

Consult the updated list of treatment devices approved by the Ministries of Health and Ecology.

In the case of approved sectors, a distinction is made between:

Compact filters: as with traditional systems, primary treatment is provided by an all-water tank. The secondary treatment is carried out within a filter bed, which can be made up of different materials (zeolite, coconut shavings, rock wool, sand, etc.) on which the purifying bacteria are fixed.

Micro-stations: primary treatment is provided by a primary decanter. The secondary treatment is carried out by the bacteria present in the biological reactor, to which dissolved oxygen is supplied. These bacteria can be fixed on support (micro-stations with fixed cultures) or free (micro-stations with free cultures). The sludge thus produced is then separated from the wastewater treated by a clarifier and returned to the primary decanter where it is stored.

Filters planted with plants (mainly reeds): wastewater treatment is ensured both by mechanical filtration and biological degradation by the bacteria that grow on the filters. They may or may not be preceded by an all-water tank.

Dry toilets

There is another type of sector which is a special case: dry toilets. In this case, the toilets are not supplied with water, and the urine and feces are collected and treated by composting, after adding or not adding wood chips or sawdust. This system must be supplemented by a treatment system for gray water.

Each of these channels has advantages and disadvantages

The traditional sectors generally do not require electromechanical elements and do not consume electricity: the oxygen supply is done naturally and, unless the topography of the ground does not allow it, the wastewater circulates by gravity flow in the works (no pump required). On the other hand, many approved supply chains, and in particular micro-stations, operate more complex and require electromechanical elements: recirculation pump, air generator for the supply of oxygen, etc.

Therefore, some cannot work intermittently and cannot be used for older homes. But apart from the filters planted with plants, approved channels have the advantage of being very compact, which allows them to be used for cases where space constraints are important and where traditional channels cannot be established. However, it should be remembered that the evacuation of treated wastewater must be done by infiltration as a priority: unless a study shows that this is technically impossible (soil that is not very permeable for example), it is, therefore, necessary to provide an infiltration system at the outlet an approved industry.

5(c). Example septic tank installation quote: choice and prices

Sanitation

Do you need to plan the installation of a septic tank? Different procedures will be imposed on you, and certain costs will be grafted so that the layout complies with the laws and regulations in force. Discover the essential steps of installing an all-water septic tank by carefully reading the estimate for installing a septic tank produced by our partner craftsman. Services and prices are notified.

The installation of an all-water septic tank, rather than the installation of another approved system such as a micro-treatment plant or compact filters, is decided according to the terrain, objectives, and budget. A soil study and then a septic tank installation diagram are carried out so that the Public Service of Collective Sanitation can study the project and give its approval. The pit installation diagram is produced by the company in charge of the works, by the architect or a design office. The septic tank receives aerobic or anaerobic bacteria depending on the system chosen.

Installation of secondary septic tank ventilation may be necessary. This device ensures the proper functioning of the system. Placed behind the pit, its role is to evacuate the gases produced thanks to a static or wind evacuation circuit. The wind turbine pipeline, which must be installed during the installation of the secondary ventilation for the septic tank, is located about 50 centimeters above the roof. The air outlet must be sufficiently distant from that of the primary ventilation. This secondary ventilation installation for septic tanks is planned from the start of the project and added to the tank installation diagram.

On average, the cost of a spreading pit is $4,500, with a price range from $3,000 to $6,000.

The price of a sandpit varies from $4,000 to $7,500,

That of an ecological septic tank from $5,400 to $11,000,

That of a single polyethylene pit from $450 to $1,200.

Example of a septic tank installation quote: budgets

The amount recorded for the installation of an all-water septic tank depends on the capacity of the tank.

The price of a 4,000-liter septic tank for six main parts is between $ 800 and $ 1,200, excluding installation.

An anaerobic system costs between $ 3,500 and $ 7,500.

An aerobic system costs between $ 10,000 and $ 20,000.

The septic tank installation quote template covers all the stages of setting up a 4,000-liter all-water septic tank, including a Zeolite filter, 25m² of the vertical flow filter bed, and a degreasing tank.

Four stages stand out: the soil study, and the passage of the PUBLIC SERVICE OF COLLECTIVE SANITATION, the earthworks, the installation of the septic tank, and the proposal for a maintenance contract.

The total amount entered on the example of a septic tank installation estimate is $8,100.00 excl.

The budget is included in the average price of ecological septic tanks. It is increased by the installation work.

Do you want to install a septic tank?

Example quote for septic tank installation, step by step

Step 1 – Soil study

The soil study carried out by a geologist specifies the characteristics of the terrain to select the ideal system. It also determines the humidity level in the soil and the possible presence of a water table. This study is requested by PUBLIC SERVICE OF COLLECTIVE SANITATION in the file before installation. A site visit is planned here. If it takes place before the works, its amount can be considered as a royalty, not subject to VAT. After the works, the visit is taxed at the normal rate.

In the estimate for the installation of an all-water septic tank, the price of the soil study and the passage of SPANC is $ 600.00 excl.

Step 2 – Earthworks

The earthwork prepares the ground for receiving the pit. A hole larger than the pit must be dug. Sand or gravel is installed at the bottom to a thickness of about 12 cm.

The cost of earthworks is included in the price of the supply and installation of the all-water tank.

Step 3 – Supply and installation of the tank

The example of an estimate for "installation of a septic tank " figures a 4,000-liter tank with Zeolite filter, 25 m² of vertical flow filter bed, and degreasing tank. The Zeolite filter is designed for the treatment of wastewater. It limits the number of oil changes required over time.

The filter bed is a natural and ecological system that avoids the need for electricity or electromechanics. No adjustment is necessary. Maintenance is, therefore, reduced. The grease trap retains oils, solid waste, and grease so that these materials do not clog up the evacuation pipes or the pipes.

Established on the septic tank installation d is, the price of the 4,000 all-water tank is 7,500.00 $ excl.

Step 4 – Backfilling

After the installation of the pit, the land is backfilled. The trenches are blocked. The all-water tank is covered with all possible precautions when it is a tank with an anaerobic system.

Step 5 – Establishment of a maintenance contract

The service or maintenance contract saves on the cost of emptying the pit. The frequency of the operation depends on the number of occupants of the house and the capacity of the tank.

Specified in the estimate for the installation of an all-water septic tank, the price for emptying the tank is approximately $240 every four years.

Example quote for septic tank installation:

The rate retained for the installation of a septic tank for new construction is the rate of 20%, as in this example of the estimate of installation of a septic tank online. This rate also applies to any installation intended for a building other than for residential use.

The rate applied for the rehabilitation of an all-water tank, and non-collective sanitation is the rate of 10%, provided that the house concerned has been built for more than two years. The installation must, therefore, already exist. This rate can be applied for all categories of non-collective sanitation, i.e., for the rehabilitation of septic tanks, micro-purification plants, and compact filters.

6.

DEALING WITH GENERAL CONTRACTORS: WHAT TO DO?

Autonomous/off-grid/greenhouse living is now more than a trend; it is a building method adopted by many families in various parts of the world. Many construction companies offer you advice and help by offering guidance and advice, which covers each of the planning stages of a new house construction project.

In any event, when environmental protection/friendliness and energy conservation are not among our needs, everybody has an enthusiasm for considering "green," for constructing their house.

Natural, "green" or autonomous with supportable improvement names are presently mainstream in every consumer market, and the building business is no special case. These ideas additionally significantly affect demand, where a few players in the home-building industry apply an enormous number of assets to their improvement and advancement.

For instance, the Canadian Mortgage & Housing Co. (CMHC) has actualized its Equilibrium home program that is right now exhibiting green/autonomous homes structured by five essential criteria: a solid, reasonable, biological, energy-conserving, productive and more affordable to live and keep up. In Quebec, the Provincial Association of Home Builders (APCHQ) has additionally begun a discussion procedure with its individuals to propose a methodology for building a green/off-grid house.

The Design Of an Off-Grid House

Engineering firms are striving hard for the advancement of this sort of development and can provide the solutions nearly as diverse as the requirements of clients. To explain your decisions in your solid and autonomous development venture, many construction companies offer you a report in three phases following the way toward building up an eco-friendly home considering the real factors about the needs of the market and the atmosphere. The house filling in as the reason for consideration of the report in hand is a contemporary style model created as a team with understudies in the latest architecture technology and extensive planning.

Ecological/sustainable or energy-efficient?

Off-grid/sustainable development, ecological house, productive vitality structure, etc., does every one of these articulations have a similar significance? Allows simply state that they speak to various parts of a bigger idea: green structure. Contingent upon the source counseled and the objective demographic, this definition incorporates a more prominent or lesser number of components that outline the administration of the systems and assets utilized and their ecological, financial, and human effects in the development of a house or some other structure.

The CMHC definition referenced above has five components, while that proposed by the National Association of American Home Builders (NAHB) remembers 7 for its guidelines for businesspeople (Green/self-governing home structure rules). In the last case, we start with the natural improvement of the locales, the productive administration of assets (materials), the proficient administration of the energies utilized, the effective administration of water, the nature of the indoor condition, the effectiveness of activity and upkeep of the development and the general effect of it on its environmental surroundings.

Right now, frequently referred to reference in off-grid structure is, without a doubt, the LEED program (Leadership in Energy and Environmental Design) created in the United States and now adjusted to different markets around the world. This program structured as a characterization instrument for green/off-grid structures for the merchandising is likewise utilized as a benchmark in private residences and incorporates five classes of criteria: ecological site improvement, effectively overseeing water resources(rainwater), vitality and environment, materials and assets, just as nature of inside conditions.

Regardless of whether you utilize 4, 5, or 7 criteria to characterize a green/independent house, the primary concern is to see how you can apply these ideas to the development of your home. This is the thing that you will find right now. You will have the option to settle on educated options about the components to organize dependent on your own ecological, monetary, and human interests.

Let us first understand the meaning of green/self-ruling structure just as a synopsis of the components to consider when you start your planning procedure, and during your first meeting with your compositional creator/architecture. We will likewise consider the components to be regarded all through the acknowledgment of a green/independent house.

Invaluable development

As indicated by Philippe Mercure, head of marketing administrations at an organization represented considerable authority in such development arranging, structuring, and usage, "green/independent development offers a large number of alternatives which, at last, should regard your preferences and your spending limit, yet it must be recalled that the advantages of such development are not just natural. The tenants of these homes straightforwardly advantage from the numerous advantages that accompany it and that regularly merit venture and exertion. "

Here is a portion of the advantages that green/self-ruling property holders can, by and large, exploit more than regular mortgage holders.

Lower support and working expenses

The significance given to vitality proficiency and sparing assets will lessen the requirement for vitality and materials. Vitality and upkeep bills will be decreased in the medium and long haul.

More noteworthy solace

Green/self-ruling houses profit by an increasingly steady temperature and a controlled environment. This likewise brings about better control of the mugginess level, and temperature contrasts are limited for better solace.

A superior quality condition

By giving specific consideration to development subtleties, the materials picked, and the mechanical frameworks utilized (air treatment and others), a green/self-governing house will contain less destructive synthetic concoctions and will give a more beneficial indoor condition.

Initial step: the site

For green/self-governing development, concerning some other development venture, the main basic advance in your methodology is surely the decision of your neighborhood and your territory. What's more, if you have at least enthusiasm for green/independent structure, it will be basic to pick an area and a land that will regard certain fundamental components.

A few criteria must be considered to pick a position of living arrangement for you and your family. Past lodging, you should likewise eat, work, and have a ton of fun. It will accordingly be important to pick an area encouraging travel by walking, by bike, and by open vehicle. The parks, schools, shops, and relaxation places should be in the close vicinity. By constraining your outings, you will decrease your green/autonomous house gas emanations brought about by mechanized vehicles by a similar amount.

Urban Planning

It is as yet hard to limit the effect of the development of a house on nature. Sylvain Charette, the proficient technologist for the Drawings Drummond office in Longueuil, clarifies that, most importantly, "construction is a damaging operation. It is important to clear the ground, burrow an opening, and adjust the widely varied vegetation present on the site. The improvement of framework and administrations causes significantly more harm when building up another road, another locale, and the assembling and transport of materials additionally produce contamination. Then again, if the utilization of such material or such a strategy permits a foundation or a structure to last any longer, truth be told, we will spare vitality.

The mission for the perfect land in a picked segment is likewise a strenuous undertaking, which is the reason new lodging improvements are progressively far off from the focuses. Nonetheless, from an urban arranging perspective, it is completely important to densify the land still accessible to control urban spread. The valuation of old mechanical land is a superb road to abstain from creating virgin land and to alter the zoning of green or farming spaces.

Development is additionally simpler in evolved parts since administrations and foundation are as of now present there. The structure of the road format ought to likewise consider winning breezes and daylight to make productive foundations and to boost their consequences for lodging. The living condition will be even more wonderful, and the avenues will have the option to be fitted out in a protected way and to advance social collaborations.

Direction and execution

When the land has been picked, it is currently important to find your home there in an ideal way. Augmenting the advantages of the sun is generally simple when the structure is very much situated. Misusing the south essence of the house permits you to exploit the light and warmth of the sun. The latent lighting and warming gave by the sun are fundamental components of your off-network development, and their effectiveness relies essentially upon this direction given to the development.

In like manner, a carport based on the north side will fill in as an obstruction against cold and wind, while coated openings toward the south will give warmth and light in the living spaces. In winter, the sun is lower, and it will infiltrate further into the house. In summer, articulated roof and shades will forestall the sun, higher in the sky, from going into and overheating the house.

In the undertaking portrayed here, we have a perfect circumstance where it is conceivable to situate the living spaces toward the south. It isn't constantly conceivable to pick the ideal land dependent on this significant basis; however, you should, at present, give exceptional consideration to it since a few components of your biological development will rely upon it.

When arranging the advancement of the land and the foundation of the house on it, and to augment the vitality effectiveness of an off-grid house in a cold and calm atmosphere like that of Quebec, we should completely take care to capitalize on the warming impact of the sun's beams in winter and make adequate shade in summer. The impact of the breeze ought to likewise be limited in winter and still permit cooling air developments in summer. To accomplish this, the area and direction of the house are key components here, which should anyway be joined with fitting finishing.

Give specific consideration to the plant species present on your property and just clear the space important for development while securing whatever number of neighborhood trees and plants as could reasonably be expected. The trees will fill in as a boundary against wind and cold in winter while hardening the environmental factors of the house by making concealed territories in summer.

The utilization of tree species and plants explicit to your property will likewise guarantee simple support and arranging, setting aside time and cash, in short just as long haul. In any case, the plants and blossoms from another condition will require substantially more watering, stuffing and upkeep time.

Sun Based Energy

Sun based energy supply structure is fundamental to the off-grid house. By just windowing the south side of a habitation, its inside will profit by the free warmth of the sun in winter. An extraordinary method to exploit this vitality is to utilize a decent warm mass before enormous windows that will store sun-oriented warmth during the day to transmit it around evening time. This mass ought to in a perfect world be on the ground since this is the place the daylight is most grounded. Solid, thick earthenware and common stone are acceptable decisions.

Adequate fenestration toward the south will permit the most extreme introduction of this warm mass in winter. Then again, in summer, it should be shielded from the sun by an overhang or an adaptable and mechanized sunshade gadget. This will keep the structure from overheating in summer and limit the requirement for cooling while at the same time diminishing visual glare.

Characteristic lighting ought to along these lines be concentrated to abstain from overheating and glare. In the field, the fundamental methodology is to plant verdant trees before the windows toward the south, while in winter, the nonattendance of leaves will empower the entry of daylight.

Structures developed with winning breezes and the course of the sun will fundamentally be increasingly agreeable and proficient.

To exploit the sun's free vitality much more viably, it is conceivable to introduce a photovoltaic framework and sun-based boards to produce power. The vitality delivered, if not quickly required, can be put away in collectors or batteries for some time in the future. It can even be sent to the neighborhood power system and offer credits to the proprietor.

Inside format

Since your house is your primary living condition, the format of inside spaces should initially relate to your requirements and your way of life. The environmental methodology, notwithstanding, needs to restrain the unnecessary spaces that require support and create extra expenses. It is, in this way, not just an issue of picking solid materials that are not destructive to the earth, yet besides utilizing as not many as could be allowed.

A thorough examination of your needs as far as several rooms and their regions is basic at this phase of arranging. Your originator planner ought to have the option to encourage you satisfactorily to arrive. In addition to other things, we will support the advancement of multipurpose rooms. Likewise, adequate mechanical space must be given to introduce the different frameworks effectively essential for your development, notwithstanding the extra innovations that can be incorporated into your off-grid house (power gatherers, dim water authorities.

There is no doubt here of denying yourself of the spaces that you think about basic; however, a decent structure of an off-grid house ought to permit you to diminish the utilization of assets. The ideal approach to do this is to boost the utilization of room. If you need four rooms to suit your family, you will fundamentally need to consider; however, you ought to likewise arrive by possessing a littler territory than in a regular house. A decent method is to survey the components of houses previously settled in your general vicinity and attempt to improve the situation for your development.

The principal living spaces will be fitted out toward the south of the house to exploit regular lighting. The model that we are introducing here incorporates a three-story format that endeavors total fenestration on its three stories on account of a slanting parcel.

A sound house should likewise deal with its occupants, so you have to know whether a two-story house is appropriate for your way of life and your capacities. Individuals with decreased portability will profit by preferring a home model, and regardless of whether this isn't your case, it is constantly imperative to consider the future if you need to live in your home for a long time.

The beginning of construction

The following stage concerns the selection of materials before development starts. A vitality proficient structure envelope and practical and solid materials will guarantee tenants exploit the advantages of off-grid development.

It's anything but a simple undertaking to plan a biological house and meet every one of its necessities. We should cause bargains, to organize our requirements and attempt to get ready for the future to evaluate their long-haul impacts. This house won't be a research facility and particularly not modern. Then again, it will be sound for its tenants, environmental to protect the planet and feasible to limit our biological impression.

An envelope with high biological potential

Progressively worried about their wellbeing and the earth, Quebecers have changed the recognition they have of their home as of late. They need to live in more beneficial, greener, less vitality expending, and increasingly manageable homes to moderate assets and what's to come.

Off-grid development starts with the decision of land and structural plans that consider numerous criteria. The association of rooms to augment space and diminish the utilization of assets is as significant as the direction of the house, as indicated by the idea of the land, the daylight, and winning breezes.

Development of an off-grid house

Right now, the group of Dessins Drummond Inc. will assist you with choosing the materials that will be utilized to make the casing and the envelope of your home. Review that the principal target of the off-grid building is to safeguard the assets of our planet while giving a more beneficial condition to its occupants. Thus, we will support recyclable materials with a base environmental impression and the greatest life expectancy.

As Sylvain Charrette, an expert technologist at the Drawings Drummond office in Longueuil, brings up, "solid and manageable lodging must, by definition, have a long life expectancy; the more drawn out the timeframe before new regular assets are looked to supplant its parts, the more natural it will be. "

Development of building envelope

It is where the best measure of materials with environmental potential can be utilized. Before beginning development, a few decisions must be made concerning the materials and methods utilized. Here are the fundamental parts of certain proposals.

Establishments

Even though the creation of Portland concrete is very contaminating (around 5% of world creation of carbon dioxide), a great solid establishment is fundamental since it has a boundless life expectancy and will permit the structure to cross the years without moving. Today, there are new sorts of cement known as "Ecosmart" available, in which a level of Portland concrete is supplanted by different less dirtying concrete increases (debris or other).

Similarly, one of the perspectives not to be neglected when raising establishments is the decision of a formwork specialist liberated from hydrocarbons. To encourage the formwork stripping, most farmworkers cover the pressed wood parts of their formwork with engine oil. In any case, since this oil is utilized on the two sides of the establishment, hints of it tend to be found in the storm cellar for an extensive period. Today, water-based, low-discharge formwork discharge specialists exist.

Furthermore, to build the solace in the cellar and the degree of general protection of the house, it is prescribed to introduce a 2-inch expelled polystyrene on the ground before pouring the solid piece from the floor.

Envelope and Frame

The edge is usually made of 2"x6" wood pieces. Wood is a sustainable asset; however, we should energize mindful utilization of our backwoods. The FSC Canada association (Forest Stewardship Council, www.fsccanada.org) proposes continually searching for items (wood, furniture, paper, etc.) bearing the FSC logo, which guarantees that the backwoods of the birthplace of the item has been overseen mindfully for nature and the human networks that rely upon it.

Another fundamental principle of the off-grid building is the conservation of assets by restricting waste. Customary development involves 3% to 5% misfortune. In an off-grid building, one must take the fundamental measures to limit the number of materials dismissed. Right now, Société habitation du Québec specifies that "the industrialization of development has in numerous regards enduring points of interest."

The utilization of manufacturing plant made segments is along these lines a brilliant "reasonable" elective that offers two significant points of interest. Most importantly, it spares assets on account of the ideal administration of materials changed in the processing plant. Likewise, works did in the industrial facility under controlled conditions make it conceivable to keep up the brilliant nature of getting together, which improves the life expectancy of the structure.

Some essential components, for example, rooftop supports and floor joists, are as of now pre-assembled generally. The construction business for these parts, which reuse their buildups and even those discharged from other mechanical production systems, is a superb case of a mindful asset to the executives.

Likewise, we could utilize built wood whose exhibition is considerably more dependable than conventional bits of wood. Despite the higher buy cost, more noteworthy utilization of designed wood would spare large measures of materials.

Insulation

Most definitely, splashed polyurethane is a profitable natural arrangement outside the structure just as inside the establishments. The item picked should comprise of reused plastic, oil, and soy buildup. This protection will, without a doubt, establish a sound, effective, and tough structure envelope.

Stone fleece is additionally a brilliant protection choice. Rock fleece is produced using basalt (a volcanic stone), and notwithstanding being a brilliantly warm and acoustic protector, it is non-ignitable, water-repellent, and penetrable to water fume. Furthermore, its strands are viewed as less hurtful to wellbeing, since they have a breadth bigger than that of fiberglass.

Another choice to fill the cavities between the uprights of the structure and the storage rooms is cellulose fiber. It is 100% characteristic and contains 80% fiber from reused newsprint. It is additionally effective from a warm perspective and penetrates the little interstices of the structure to frame a homogeneous pad.

Regardless of whether your development is off-grid or customary, predominant warm protection is, without a doubt, the best venture you can make. Notwithstanding expanding your solace, it will diminish your vitality utilization.

Outside Coatings

For Mario Paquette, specialized chief at Dessins Drummond, the wild stone is the undisputed boss of off-grid items right now. Fieldstone doesn't require any change from the start, which isn't the situation with ashlar, which is separated and cut precisely.

The upsides of normal stone (cut or field) are various, in any event, when you think about its more significant expense. It is, as a matter of first importance, the most solid of the alternatives with a boundless future. It is likewise an impartial material that radiates no harmful substances, and which is reusable. It requires no support and opposes the most noticeably awful climate. A stone divider likewise offers prevalent warm and acoustic protection. The main disadvantage related to regular stone is the separation it should make a trip to find a good pace since its vehicle requires a great deal of fuel.

Block and made stone are additionally acceptable decisions since they offer indistinguishable preferences from stone, yet with a bigger environmental impression, be that as it may, since their production dependent on concrete is more dirtying than normal stone.

Regular wood – in a perfect world untreated, recolored, and secured with normal items – stays a legitimate decision. It is generally sturdy, simple to fix or supplant, and recyclable. T wood is additionally a great environmental alternative. Entirely strong and impervious to shape, no synthetic compounds are utilized in its change. It is anyway somewhat more requesting than common wood from a natural perspective since cooking comprises warming the wood to more than 200oC in a broiler. Along these lines broiled, the wood gets waterproof, decay evidence, and hard, which will give it a generally excellent period of usability.

Roofing Material

Albeit conventional asphalt shingles are currently progressively being recuperated and reused, the most ecologically inviting decision despite everything is by all accounts sheet steel or light steel shingles. This material is considerably sturdier and safer than black-top shingles and is completely recyclable.

It is additionally prescribed to pick pale hues, clarifies Sylvain Charrette since they will reflect more daylight, which will have the impact of overheating the storage room. Likewise, particular kinds of steel sheets won't need help, for example, pressed wood or situated strand board (OSB), which diminishes the utilization of wood.

Note anyway that arranged strand board (OSB) can likewise be utilized as a feature of a sound home since many are presently made without formaldehyde-based cover. The shavings utilized are likewise reused wood buildups that stay away from the dumping ground or the woodstove.

Windows

Concerning natural development, the fundamental capacity of windows is to permit us to profit by free and boundless vitality: sun-based vitality. You should initially pick vitality proficient windows affirmed "Vitality Star." These confirmed windows are about 40% more effective than customary windows, they limit heat misfortune while separating more UV beams, and they are increasingly sturdy.

With regards to the sort of skeleton, the most earth well-disposed and strong decision would be the fiberglass body; however, not many organizations offer this choice. Different materials either have higher biological expenses or are hard to reuse. Also, fiberglass is a magnificent protector.

An off-grid house focuses on the ideal utilization of sun-based vitality in the entirety of its structures. Plentiful and very much situated windows (south-south-east) must be joined with a smart general plan which incorporates, in addition to other things, congregations of staircases without risers to permit the beams of the sun to go notwithstanding improving the course, between the floors.

To guarantee the solace of the tenants, abstain from overheating and lessen cooling costs; it is additionally fundamental to introduce a sunshade, for example, a mechanized visually impaired or canopies, which will permit sun-powered radiation to be controlled by current needs.

Fundamental Decisions

We should likewise recall that for all the items we purchase; transportation is a component that adds a great deal to the natural impression that we leave. It is consequently constantly desirable to empower the acquisition of neighborhood items as a lot to protect our condition as our portfolio.

Building the envelope of your house is a stage that requires huge amounts of materials. It is in this manner at this phase, a few natural strategies and maintainable materials must be utilized if you wish to safeguard the assets of our planet while giving a superior quality condition to your family. When the structure and envelope of the house are finished, we will, at that point, continue to the establishment of the mechanical frameworks and afterward to the inside wrapping up.

The inside completion

The selection of materials is fundamental to establish a solid inside for the soundness of the inhabitants. Specifically, we should attempt to dispense with poisonous items and materials. Unpredictable natural mixes (VOCs) discharged from molecule board, mortars, covers, paints, varnishes, and glues are much of the time the reason for sensitivities and respiratory issues in easily affected individuals. In a perfect world, one ought to ask about the creation of every material and dispose of solvents and materials containing formaldehyde.

Ideally, use sans VOC paint, water-based types of cement, and inherent strong wood or particleboard without formaldehyde. The ground surface will be made of hardwood from a very much oversaw timberland (FSC), bamboo, plug, cover, fired, or regular stone introduced with water-based glue.

Ventilation and warming

To have a really strong and durable home, ventilation must be powerful. In an overly tight house, you should introduce a mechanical ventilation framework with heat recuperation (HRV). Albeit progressively practical to purchase, a straightforward air exchanger isn't sufficient. An HRV devours almost no power, and it recuperates in any event 80% of the warmth from the fumes air. A few models additionally permit dampness to be gathered to abstain from drying out the air in the house.

Also, the circulation of outside air must be guaranteed in all rooms and the parlor. Stale air will be evacuated via air admissions in the kitchen and restrooms. Appropriately adjusted ventilation ensures solid indoor air for its inhabitants.

Since it permits investment funds of 65% on the warming bill, geothermal vitality is the most natural method of warming. Therefore, the productivity of geothermal vitality is acquired at the expense of enormous speculation, which is hard to legitimize with regards to off-grid development. Also, in Quebec, introducing an electric warming framework (constrained air or baseboards) stays an ecologically mindful decision. On account of baseboards, indoor electronic regulators will permit investment funds of around 10%.

Air Flow

Air Flow can lessen your heating bill. A decent air snugness will boost the protection of warm vitality. Just Novoclimat homes have a snugness rate checked by an infiltration test. A Novoclimat house will spare you at any rate 25% of your vitality charge contrasted with standard development.

Energy Star image

When buying family unit apparatuses, hardware, entryways and windows, and warming, ventilating and cooling gear, consistently scan for the Energy Star image. Gadgets set apart with Energy Star don't cost more, and they will set aside your cash while securing the earth. Note that Energy Star entryways and windows are obligatory in Novoclimat homes.

Water utilization

The reasonable utilization of water is likewise part of a biological house. Sterile offices with low water utilization will be introduced, for example, double flush toilets (3L and 6L), just as temperature and stream controllers for the shower and the bath. What's more, a water gathering framework ought to be utilized to water the vegetable fix and finishing plants.

It's anything but a simple errand to fabricate a biological house and meet every one of its necessities. We should bargain and make a point to give a second life to the deposits and waste during the erection of the house. Our homes must be sound for inhabitants and manageable to limit our environmental impression on the planet.

7

DESIGNING AND PLANNING OF AN ECOLOGICAL/OFF-GRID HOME

7 (A) Eco-friendly house: save money while respecting the environment!

The ecological house, or off-grid building, is no longer reserved for experienced ecologists: more and more builders and architects specialize in this buoyant sector, which promises significant energy savings while ensuring respect for the environment. An ecological house is a house designed to be respectful of the environment. This type of habitat must, therefore, create as little pollution as possible while reducing energy needs and losses.

For this, we will intervene in different aspects:

- the design: the plans for an ecological house must be carried out by ensuring that it is in harmony with its environment;

- building materials: to build an ecological house, we will favor the use of materials of natural origin, recyclable or materials that do not produce polluting energy;

- equipment: it must make it possible to reduce energy consumption by using alternative heating or domestic hot water production systems.

An ecological house should ideally respect the following principles to be considered as such:

- **environment-friendly site**: the site must use as little energy as possible while minimizing waste.

- **ecological building materials:** we favor materials of natural or recycled origin, non-toxic, and, if possible, locally produced (wood, terracotta or raw bricks, straw, etc.).

- **Take advantage of the environment and the climate:** optimization of the sunshine in the living rooms, protection against the wind by planting trees, installation of double or triple glazing - possibly with solar or electrochromic control – or the achievement of a roof to limit the sun's rays in certain rooms.

- **optimal thermal insulation and ventilation**: use of insulating materials of natural origin (linen wool, hemp wool, wood fiber, cellulose wadding, etc.), choice of the type of windows, and installation of a system of ventilation allowing both the maintenance of a constant temperature and the renewal of the air in the different rooms.

Have an architect accompany you! To help you make your choices and optimize the energy performance of your future off-grid building, we recommend that you be accompanied by an architect. The latter will help you take advantage of climatic conditions, inform you of the latest developments in energy savings, advise you on insulation, ventilation, and humidity management, and allow you to give a unique style to your ecological home!

7(b). What equipment should be favored for off-grid building?

Choosing an eco-construction implies having recourse to installations and equipment adapted to this type of house:

Ventilation of ecological dwelling ventilation of an ecological dwelling

In a house, ecological or not, the ambient air must be regularly renewed to preserve the health of its occupants and avoid problems related to humidity (condensation and mold, in particular). If natural ventilation is possible, it is often recommended and more efficient to install mechanical ventilation. We will favor the installation of a double-flow VMC, which greatly limits heat loss while using the energy extracted from the indoor air to heat the incoming air. In addition to filtering outdoor pollutants, this type of installation allows savings of up to 15% on your heating bill. You can also opt for a humidity-adjustable single-flow CMV, more economical but less efficient, which will increase its suction flow when the indoor humidity level exceeds a certain threshold.

Heating and domestic hot water

Have an efficient ecological heating system installed, which will allow you to use renewable energies to heat your interior and produce hot water for domestic use.

Thus, the air-water heating pump makes it possible to produce both heat and domestic hot water, using the energy of the outside air and a little electricity. This system works independently, allows you to maintain a constant temperature in your home, and divide the heating bill by three. A heat pump of this type only works with low-temperature radiators and/or a heated floor.

You can also opt for a wood stove, which provides constant and gentle heat. Preferably choose a wood pellet stove, because their performance is more important than that of logs, and they are fuels from wood waste (sawdust), therefore, it is more ecological. Note that some wood stoves also produce hot water.

Another possible solution is solar heating. Thermal collectors installed on the roof of the house store solar energy and transmit it to a water heater and/or a heated floor. If this system only uses solar energy to operate, it must nevertheless be combined with a wood stove to be able to operate continuously. Despite this, solar heating can reduce the heating bill by 50%.

Electricity in the ecological house

Electricity: produce it yourself!

When you have the project to build an ecological house, producing your electricity becomes obvious.

The installation of solar panels can be an interesting solution if one of the slopes of your roof has good sunshine, an inclination of 30 °, and is rather exposed South. Know that aero-voltaic is four times more efficient than photovoltaic because it also recovers the heat produced by the panels to heat your interior. It can also generate fresh air during summer nights.

On the other hand, the wind turbine is not as profitable as a means to produce its energy: it costs 2 to 3 times more expensive than photovoltaic panels and needs to be located in an area exposed to strong and regular winds to produce enough energy.

Finally, a cogeneration boiler costs 2 to 3 times more expensive than a conventional boiler but offers better performance and also generates electricity (50 to 80% of the electricity needs of a home).

As the energy generated most often depends on natural elements (sun, wind, etc.), keep in mind that this energy cannot be produced continuously and that it is, therefore, necessary to supplement this supply with "public" electricity. If you wish, you can resell all or part of your electricity production to EDF to amortize your costs more quickly.

Low energy bulbs to save energy

Lighting: opt for LED bulbs. Lighting represents up to 10% of the electrical consumption of a home. With prices that have been falling for several years and great energy savings to be had, low-consumption bulbs have everything to seduce! Indeed, if their cost remains high enough, it is offset by a lifespan of up to 25 years and an energy bill up to 10 times lower than with conventional bulbs! Among the low-consumption bulbs, LEDs are the ones that offer the longest life and the lowest energy consumption. Indeed, an LED bulb or spot will have longevity six times longer than a compact fluorescent bulb.

Save water in eco-construction

Water consumption can be greatly reduced by installing certain devices. Think of equipping your taps with regulating-saving water tips, install an economical showerhead, equip your toilets with 2-speed flushes or install dry toilets, and set up a water recovery system. Rainwater can be collected and filtered to be used for toilets, soil maintenance, washing clothes, or used without the filter to water the garden or clean your car. Also, by choosing a dishwasher or a washing machine of class A + or higher, you can save up to 50% of water compared to the old models.

Household appliances: check the energy class

If you want to buy traditional household appliances, make sure they are energy efficient. For this, preferably choose devices of energy class A + at least, which guarantee optimal energy consumption. This codification, which classifies the devices on a scale from G (poor efficiency) to A +++ (optimal efficiency), takes into account many criteria: electricity, water consumption, lifetime, noise, capacity, etc.

There are also devices operating on renewable energies: wood stove, solar oven, etc.

Estimate the price of your future off-grid home!

7(c). Choice of materials for the construction of an ecological house?

To build an eco-friendly house, it is important to favor natural and sustainable building materials. Here are the most common materials used in the construction of a detached home:

Wood

It is impossible to speak of an ecological house without mentioning the wooden house! Indeed, for many, wood represents the most suitable material for such a construction. If for the moment there are only 7% of wooden houses in France, this material is more and more popular: wooden constructions have increased by 15% per year for several years. Wood is a recyclable and renewable material, which has the distinction of being 15 times more insulating than concrete. It is most often used to construct the framework of buildings, which is then filled with other ecological materials, such as hemp or straw.

Hemp

Hemp is the ecological material par excellence: this plant grows very easily, requires little fertilizer, little water, and does not require pesticides. In construction, hemp is mixed with lime to form a kind of concrete. This "concrete" is then placed between formwork boards, the formwork, which is removed once the wall is dry. This wall thus formed is then protected with a coating, both outside and inside. The design of hemp walls requires certain expertise. Ask a project manager!

Straw

Building a straw house can seem quite utopian. However, this type of ecological construction is as solid as the others! The principle is rather simple: the frame of the wooden construction is filled with straw bales, which are then coated with lime. The thickness of the walls thus obtained allows a very good insulation capacity for the house. If this idea tempts you, the straw being considered as an insulator, your builder cannot offer you the usual 10-year guarantee on your construction. It is much more durable and economical than that.

Monomer brick for green/detached building

Monomer bricks are very thick terracotta bricks with alveoli. They have built-in air and high grade of thermal inertia that isolate the construction. These bricks are very resistant and allow construction on several floors. They are easily used by professionals who adapt to them without a problem. However, monomer bricks are not perfectly ecological: their manufacture, and especially their firing in gas ovens, requires a lot of gray energy!

Raw earth house

The mud has several advantages: it is easily available, regulates the humidity of the ambient air, reduces heat exchanges between the inside and outside (it retains heat in winter and keeps cool in summer), and its manufacture requires very little energy (it does not need to be cooked). Its installation requires using the same technique as for hemp walls but must be completed by a stone or brick base, and roof overhangs to protect the house from bad weather.

7(d). The different types of ecological houses or eco-constructions

The BBC house

The BBC house (Low Consumption Building) is the most widespread ecological house today in France. The BBC label aims to set a limit not to be exceeded in terms of energy consumption for new constructions. Houses that meet this standard must then comply with certain measures, particularly concerning the five primary uses of a home: lighting, heating, air-conditioning, hot water, and ventilation.

The maximum limit of energy consumption not to be exceeded then depends on the location of the building (altitude, geographic area, etc.) and the living area. On average, a new house complying with BBC standards should not exceed 50 kWh of primary energy/sq.m/year.

The BBC standard has been made compulsory for all building permits since January 1, 2013, as part of the new thermal regulation for buildings, RT 2012. This type of housing is 3 to 4 times less energy-consuming than 'a house built according to previous standards (RT 2005).

The bioclimatic house

A bioclimatic house's principle is to use the natural resources of its location. The principles of this ecological habitat are notably used within the framework of construction with high environmental quality (HQE) approach. It then uses common sense principles for a house that consumes a lot of energy and is pleasant to live in.

Here are some basic rules for a bioclimatic home:

- orient the house in the south to heat it naturally at a lower cost;

- use deciduous trees to provide shade for living areas in summer;

- install the living rooms in the south and reserve the orientation to the north for rooms that can stay cooler.

The bioclimatic house seeks to replace certain heating systems, which can be very expensive, such as solar collectors or heat pumps, by favoring building materials allowing very effective insulation of the habitat. The materials used help to stabilize temperatures, to store heat in winter, and to keep it cool on summer days.

The passive-house

It is the type of house requiring low energy usage, responds to the following principle: the heat given off by the interior of the house, as well by the appliances and equipment as by the inhabitants, and that brought by the sun outside, is sufficient to heat the building. The building rules are then the same as for a bioclimatic house, with a large space left for very high-performance insulation. Added to this is a significant amount of sunshine on the rooms facing south, with glazing, which then represents between 40 and 60% of these surfaces.

This type of eco-friendly house is particularly present in the

Nordic countries, but also Germany and Switzerland. In France, a construction cost of at least 20% more must be expected than for conventional construction.

The passive house, a type of ecological house

The positive house or positive energy building aims to produce more energy than it consumes. This type of house, self-sufficient in energy, uses the same construction principles as a passive house but is also equipped with an energy production unit, such as solar collectors, a heat pump, a wood boiler. These houses require significant financial investments at the start, but are profitable after a few years, especially if you sell part of the energy produced to EDF.

7(e). What price for an ecological house?

The cost of an ecological house is, on average, 10 to 20% higher than that of conventional construction. However, the price of green/detached buildings must be put into perspective: after a few years, the energy savings made will compensate for the additional initial cost, and such a house will be sold much more easily than a traditional house.

Average price per m² of a green/detached house

Type of ecological house Price per m² (*)

BBC house 1,200 to 1,800 $ /m²

Bioclimatic house 1,500 to 2,500 $ m²

Passive house 1,500 to 3,500 $ /m²

Positive house 1,500 to 3,500 $ /m²

(*) The low price range corresponds to the average price of a house charged by builders of individual houses, while the high price range corresponds to the average price that you will pay through an architect.

7(f). The six points to keep in mind while constructing

Because improvising is not an option, go from dream to reality without pitfalls or unpleasant surprises, with the experts of Ecohabitation.

You have the crazy project, but oh how inspiring of an ecological house? Building a new home is truly quite a business, but if it is planned properly, the project is much easier to achieve. You will be ready to face the unexpected! Ecohabitation gives you six tips to follow absolutely to make your project a success!

1. Make a realistic budget and schedule

Plan, plan, and plan again. This is lesson number for a successful project! It is imperative to make a pragmatic budget and project schedule. Take into account all the items of expenditure and spread your budget according to the schedule, to be able to correctly forecast the expenses and avoid surprises.

It is also very important to plan the additional costs; "from experience, a good margin of error, up to 20%, is the rule to prevent it from turning into a nightmare," explains Emmanuel Cosgrove,

CEO of Ecohabitation. By the way, this also applies to renovation projects!

2. Take the time to save

Building yourself takes money, lots of money. Before diving, make sure you have saved up to your project and have enough funding. If you are a tenant, be sure your back is strong enough not to be caught short of funds mid-project.

Do not act urgently!

3. Bet on a reasonable area appropriate to your needs

Do you have manual skills and like to handle the saw and the hammer? You say to yourself: "if I build myself and with all the contacts I have in the field of construction, I will save so I can afford to build bigger." Attention, this is a typical serious error!

It is wrong to believe that the option to build yourself is necessarily economical. Indeed, statistics show that Quebecers are always building bigger, too big. It is important to determine the size of your future home based on your present and future needs.

Assess your real needs and get advice from your designer or architect. Don't make the mistake of spending money on unused space.

4. Have detailed plans and material choices

This factor is decisive for the success of the ecological component of your project. Be very precise in your choices, and plan everything in detail. Surround yourself with experienced and trusted people and get informed.

Never leave room for hesitation or uncertainty with your subcontractors and contractors. Do not be limited in your vision but express it clearly and precisely. Your interlocutors do not read your mind. As these must commit to a timetable and a budget, everything must be written in black and white before being signed.

It is recommended to make a very precise list of the necessary materials, with their price and the name of the store. Don't leave shopping as a last-minute activity. Product availability is always a problem and can cause significant delays on a site. Buying for yourself can be an option, but a general contractor often benefits from preferential rates and knowledge from experience.

5. Move (only) when the interior finish is completed

All contractors will tell you, delivery delays and budget overruns are common. You may not be able to move in on the date you previously planned. Plan a buffer period to allow the site to be finalized and a smooth move.

"My lease is over, we're moving!": our team has often heard it. This is also the decision that many make, even if the project is still under construction. Before moving in, we recommend that the main living area be completely completed, up to the last coat of paint. Avoid living in-between or in a step-by-step project; " I will finish the shell, and then I will finish the kitchen, and then I will do the bathroom, etc."

"From experience, it's all but pleasant to live in a half-finished space, especially with the whole family." Emmanuel Cosgrove, director of Ecohabitation.

An incomplete project can quickly become a logistical puzzle. It is a safe bet that you will never finish the interior finish, often for lack of time or funds. It is also an important source of stress and tension. We don't want to have to finish the house in an emergency and rush the work!

6. Team up with Experts

This step is optional, but it could be the best decision for your project! Plan your project from A to Z with a team of experts, make the best choices from an ecological point of view, sustainable and adapted to your budget. Take advantage of the experience and expertise available to you.

8.

YOUR ENERGY-INDEPENDENT HOUSE

Many electricity production materials can now be installed in your home, whether you live in a house or an apartment. Several green/detached solutions are also available to households who want to be energy independent without polluting the planet. The solar panel or the photovoltaic module is undoubtedly the most popular of the solutions for producing electricity because it uses energy that is always available: the rays of the sun. The domestic wind turbine, meanwhile, is intended for houses that are well exposed to the wind. The exploitation of hydroelectric power is the ideal option for isolated buildings located on the banks of a river.

Biogas is renewable energy produced from the fermentation of biomass, household waste, and animal excrement. Individuals who want to create their production plant will then use a domestic methanizer, which they will install in their garden. The eco-generator is not, strictly speaking, a green/detached solution for producing electricity, because it exploits the natural gas of the boiler. This solution is nevertheless capable of producing heat, electricity, and domestic hot water.

What solutions are suitable for each type of accommodation?

Among all this equipment, only solar panels and the eco-generator are eligible in condominiums, the other electricity production solutions requiring more complex infrastructure that it would be impossible to install. Thus, households wishing to harness the energy of the sun can obtain small solar kits that produce enough electricity to power their light fixtures and their energy-efficient devices. Besides, if the trustee agrees, they will be able to put larger signs on the roof. They will then be able to produce more energy for heating, for example, or to power their large appliances.

Finally, an eco-generator placed at the level of the boiler of a building produces electricity for lighting common areas.

Comparison of equipment producing electricity

The various equipment used for the production of electricity each has its advantages and disadvantages. Solar panels are interesting for homes that have good solar exposure. Also, they ensure optimum performance if the roof is exposed due south and if it is inclined at 30 °. Households also have the possibility of reselling their production surplus to electricity suppliers like EDF OA, because the latter should buy electricity production from individuals at attractive rates.

As for the domestic wind turbine, it is the ideal solution for houses exposed to the wind and which have no immediate vicinity because it is noisy and can cause noise pollution, which could disturb the comfort of neighbors. Besides, this system provides a lower yield than that of photovoltaic panels because of its limited height. The overproduction of energy can also be sold to the EDF network at a very advantageous price. The latter is not, however, obliged to buy back this electricity production if the house is not located in the Wind Development Zone.

The production of hydroelectric power is ideal for housing located near a river or a torrent that benefits from a good elevation. It requires the use of a turbine, which will be activated by the flow of the watercourse. Other installations are nevertheless necessary for the collection of water and its channeling to turn the turbine. This solution is very interesting because it produces energy 24 hours a day.

Biogas is particularly interesting for farms or houses that produce enough waste to supply a domestic anaerobic digester. The organic residues will be poured into a tank, and they will ferment there for a few days to produce methane. This gas will then supply a generator with an average power of 1,500 watts. The domestic biogas installation does not take up much space on a site, because it consists of a tank of approximately two cubic meters. Finally, its use does not allow a dwelling to be entirely autonomous in electricity, because it must be combined with a photovoltaic or wind solution. Besides, the waste residue can be transformed into compost. The eco-generator, on the other hand, is connected to the natural gas network. Its advantage compared to other systems, is that it also produces heat.

Capacity and cost of installations

The cost of installations producing electricity varies according to the equipment and technologies used. Solar panels: ecological, efficient, and profitable at will, they can also generate regular income for a household for 20 years, while preserving the environment. According to specialists' estimates, a correctly oriented 1 m² solar panel will produce, on average, 100 kWh of electricity per year. On average, it is necessary to invest between $10,000 and $13,000 for panels with a power of 3 kW (peak kilowatt), between 416,000 and $19,000 if the power of the panels is 6 kW and between 25,000 and $ 35,000 for panels with 9 kW of power.

Regarding the domestic wind turbine, its power varies between 100 W and 36 kW. Its price is between $10,000 and $15,000. A household can pay up to $40,000 HT if installation costs are included. Hydroelectricity has an average power of 1,500 watts. This device is worth around 3,500 euros, while the cost of these accessories can be between 1,000 and 2,000 euros. The methanizer is capable of producing 5,000 kWh of electricity per year on average. Its price is quite substantial because it can reach 24,000 euros, but its installation can benefit from a tax credit to taxpayers. The eco-generator requires the installation of an external combustion Stirling engine, which is connected to an alternator at the level of the boiler. Its thermal power is approximately 6000 W, and its electrical power is 1000 W. Finally, its efficiency is sometimes greater than 100%. Its cost is between 10,000 and 20,000 euros.

8(a). All about solar self-consumption

The autonomous house now stands as the essential building 21st-century. Bioclimatic house, zero carbon emission housing, or ecological house, there are many terms to designate this new mode of energy production, today accessible to all. Whether from an eco-responsible perspective but also to gain independence and achieve significant savings. In conclusion, many individuals are turning to electric autonomy. How to achieve energy self-sufficiency? Is it possible to produce enough electricity to self-consume 100% throughout the year?

The habitat of tomorrow

Collecting rainwater, heating with wood, or even cooking food in the stove. These practices have long appeared as obsolete techniques, allowing only to make derisory savings. This bad image is a thing of the past. Even as current customs encourage us to gain comfort and devalue human effort in favor of the interventions of the ultra-versatile connected object, society is turning to a new model of energy consumption. Consume less to earn more and live better: this is the new collective objective. To drastically reduce its electricity bills and its carbon footprint, photovoltaic self-production, in this case, stands out as the winning solution on all fronts.

Zero CO2 emissions, zero waste, the sun offers an inexhaustible and free source of energy to meet your needs: what more could you ask for? Photovoltaic electricity is accessible to everyone and switching to solar self-consumption means choosing a responsible solution.

More savings, more autonomy, less nuclear energy: solar self-consumption is virtuous, and we will explain why.

Below, you will be able to get an answer to your following questions:

- How does it work?

- How many panels are needed?

- Is it profitable?

- How to gain autonomy?

- How does solar self-consumption work?

How does your solar installation allow you to generate electricity?

In an energy-efficient house, you produce your electricity from the sun. Update on how photovoltaics work:

The sunlight is captured by the cells of the solar panel.

Silicon, the semiconductor material that makes up solar cells, releases electrons.

The electrons move: their circulation produces an electric current.

The current, once transformed into alternating electric current by the inverter, supply your house with electricity.

High-performance solar panels adapted to your energy needs, combined with the action of the sun, is enough to produce the electricity necessary for beneficial energy autonomy. With 20 to 36 panels depending on the house, the annual production of the panels is equivalent to the annual electrical consumption.

What savings?

Photovoltaics, once (well) installed, requires no running costs, as long as the sun is free! However, it allows a significant saving on the household budget. By fueling yourself exclusively with your self-production, you no longer pay electricity bills. To a lesser extent, partial self-sufficiency in energy can greatly reduce the final amount of your supplier's invoice.

The autonomous house also offers its inhabitants new freedom: independent of a supplier, they perfectly control their energy budget. It is this feeling of independence that motivates many individuals to build or renovate their homes to gain energy independence. Tired of paying bills, eager to no longer suffer the rate hikes imposed by an electricity supplier, tired of network cuts: you have the possibility of no longer being dependent on a third party, you're controlling your energy supply. This approach goes in the direction of other societal evolutions just as fraught with consequences. By consuming locally, by using collaborative systems between individuals –

Airbnb and Uber at the top of the list – consumers express their desire to be freed from the economic monopolies of large companies. All this to save money but also for beneficial independence and minimization of constraints.

Another strong argument in favor of the so-called autonomous house, ecology. More than ever at the heart of government concerns, the issue has raised the awareness of the majority of the citizens of the planet today - for the greater good of the planet and of our children.

What is the impact of nuclear power on the ecosystem?

By using the traditional electrical network, the consumer indirectly endorses the choice of nuclear power. When France entered nuclear power in the 1960s and 1970s, it was a very good choice given the alternatives available to produce our electricity: oil, gas, and coal. Now is the time to adopt a new energy model based on renewable energy.

With a constant production of radioactive waste, the environmental impact of nuclear power is very negative compared to that of solar energy. The infrastructure necessary for the production of nuclear electricity consumes a great deal of energy, starting with the manufacture of the materials necessary for the operation of the power plants. The treatment of radioactive waste makes it possible to reduce but not eliminate radioactivity. For a long time, we have exported our radioactive waste to unscrupulous countries. While today, this is no longer possible. We must find a storage solution in our territory. Especially since some waste from used nuclear fuel has very long lifetimes exceeding one million years.

Photovoltaics is an example for the planet. Composed of silicon, solar cells are made from a particularly abundant natural resource and considered to be inexhaustible. Silicon is present in sand and rocks. Solar, to function, does not upset the balance of natural resources.

Photovoltaics as a gesture for the planet

By living in an independent house, the individual ceases to enrich the owners of power plants on the one hand. And on the other hand, it helps protect the environment. The results of the energy self-sufficiency approach are all the more likely to be successful as the number of solar installations increases. Marking its commitment to green energies is a first step, whatever the scale.

By choosing to self-generate your electricity with photovoltaic panels, you are at the heart of the energy revolution in progress. You contribute to the emergence of an energy model without environmental pollution. A symbolic but also effective act, energy autonomy is beneficial on all levels.

An already winning bet

In the past, individuals who implemented the means necessary to live in a so-called autonomous house appeared to be marginal. Often assimilated to heirs of the hippie movement of the 60s. Today, energy independence is trendy.

The house of the 21st- century is ecological. Architecture and design magazines abound. No more rudimentary buildings, the bioclimatic house shines today with its clean lines, its clever layout, and its benefits in terms of energy self-production. Aware that solar is the future of electricity, architects and designers in association with panel manufacturers draw sober or original installations, camouflaged or enhanced, but always pleasing to the eye and in harmony with the built environment.

The success of photovoltaics on a global scale is such that a famous magazine specialized in green/off-grid technologies recently announced that solar electricity will become the first source of energy by 2050. In this context of the inevitable energy transition, it is undeniable that it is necessary to bet if it is not already done, on an energetically autonomous house.

Add a new source of energy to the house

When you equip a house with a photovoltaic installation with a view to self-consumption with the sale of its surplus, you supply it with two sources of energy: the electricity produced by the solar panels and that of the electrical network to which the house remains connected.

The principle is simple: the electricity produced by solar panels has priority. Thus, it is consumed instantly on-site to fully or partially cover your energy needs. Then when these needs are more important than your solar production, the network takes over to provide the backup.

But that's not all! It also works in reverse: you too have the possibility of injecting your production (or what is left of it) on the network. So, when the production of your panels exceeds your energy needs, you can sell your surplus solar electricity to EDF Obligation to Buy for 20 years at a fixed price, and/or store it!

Let's observe a typical day: Day-to-day self-consumption curve Earth-type

7 AM: your alarm clock rings! Your phone shows a 100% charge, and you unplug it to be able to read the news of the day while enjoying a good hot coffee, like the shower you take after this one. Before leaving, you will program your dishwasher so that it starts in the morning. It is daylight, your solar panels produce and power your electrical devices.

1 PM: the sun is at its zenith, and your photovoltaic panels are producing. If you are outside, you do not benefit directly from your peak production. Your dishwasher, programmed beforehand, has just finished its work and has switched off: there is no longer any energy requirement. So, you sell your surplus solar electricity.

7 PM: The day ends, and your appetite grows, as does your energy needs! You prepare dinner on your lit work surface while watching your television from time to time. As your digestion progresses, the sun goes down, and your photovoltaic panels

produce less and less. Your energy supplier supplies you or your battery.

As you will have understood, solar self-consumption favors instant and/or stored consumption of your solar production when possible. But when it is not, the sale of surplus electricity makes it possible to financially compensate the expenses linked to the supply of the network by your usual energy supplier.

Photovoltaic and solar thermal: what difference?

Main – and essential – accessory of the autonomous house, the photovoltaic allows to self-generate its electricity for big energy self-sufficiency. Other green/off-grid technologies and certain ancestral tricks make it possible to further optimize its autonomy.

Solar water heater: zoom on the solar thermal panel

While the photovoltaic panel transforms sunlight into electricity, the solar thermal panel collects the heat from the sun to store it as such. The individual solar water heater is powered by thermal solar panels. They thus make it possible to heat the water of the autonomous house – provided that the sunning is sufficient. In no case, however, does it produce electricity.

How to articulate photovoltaic solar and thermal solar?

While photovoltaic panels make it possible to live comfortably in great autonomy, the technology of solar thermal is not sufficient.

To choose, it is, therefore, essential to favor photovoltaic solar panels. The consumer can nevertheless associate CESI and photovoltaics. The electricity produced by the panels will be devoted to the other electrical stations of the autonomous house, for an energy-saving which can prove useful.

Revolution To The End

Your historic electricity supplier may be replaced. You can buy electricity from producers of 100% renewable energy. This encourages transition and is good for local and sustainable employment. All this provided you choose the green/off-grid supplier correctly!

Solar self-consumption at home

I have good news for you! Provided that your electrical panel meets standards, no matter where you install your panels (on the roof, on the ground, or a garden shed), your household electrical network does not change!

To be able to operate in self-consumption, a photovoltaic installation is composed as follows:

- Solar panels: it will capture the irradiation of the sun to produce direct current energy.

- An inverter: it is generally placed near your electrical panel, and its role is to transform the energy captured into alternating current. The same as the one used by your current network.

This material must be selected according to its specificities to correspond to your needs and technical possibilities. A good photovoltaic installer makes it a point of honor to optimize and master all aspects of the project throughout its implementation.

To have a real vision on its consumption, you also have:

- **A smart meter**: it measures consumption. Optionally, you can also switch on certain devices when there is excess energy.

- **From an application:** available online or on a smartphone, it allows you to follow your production and consumption in real-time and then keep a history of these.

To check the capacity of your accommodation to accommodate a self-consumption installation, you can use the WattNext simulator. In seconds, it will tell you free of charge and without obligation if your project is feasible, profitable, and how convenient!

Where to install the installation?

Flexible and adaptable, a photovoltaic installation can be installed on different types of supports as long as the reception surface is large enough and has a correct exposure to sunlight.

Why, will you ask me? Because from the first rays of the day, the panels produce your electricity to power your devices (appliances, light fixtures, chargers, heating, etc.), and the longer and longer the exposure, the more you will produce. It is for this reason that the choice of the orientation of your installation is not trivial.

Note: when the weather is cloudy, the light is less important, but it is always present. The clouds act as diffusers by distributing the rays evenly. So, your solar panels are still producing!

Your photovoltaic installation can be placed on a sloping roof:

Superimposed: your panels will be fixed on rails so that your current roof is preserved. It is an efficient, fast, and economical installation.

Integrated into the frame: your panels will be installed in place of the existing roofing elements (tiles, slates, etc.) in whole or in part. It is a more complex installation and requires a longer completion time. If you are building a new house or want to renovate your roof, integration into the building saves you square meters of roofing.

If you have a flat roof, you can also have solar panels installed. These will be fixed on ballasted and oriented supports, quite simply!

The Solar Shelter: an aesthetic, practical and sustainable alternative

If you do not have a suitable surface for installing your panels, I suggest that you take an interest in the sun shelter. It is an eco-responsible wooden structure made in France and equipped with a photovoltaic roof, which we have designed for you.

The dead-end on the concrete was made for the benefit of a screw foundation, which allows it to be installed quickly and easily (in 48 hours). In addition to producing electricity, this shelter also allows you to create a solar Carport: a covered parking lot that produces solar energy. Smart!

To simulate your solar self-consumption project with a solar shelter on your property, you can use the WattNext simulator. It calculates profitability and your earnings over 25 years according to its location!

How many solar panels are needed?

On average, homes in solar self-consumption are equipped with around 8 to 25 solar panels whose unit power fluctuates around 350 Wp. But beware, the mistake would be to believe that the number of panels depends on the surface that can accommodate them. Even if it imposes limits, the first question to ask is:

"How many solar panels do you need to make your project profitable? "

Because we will agree, self-consumption is a very good idea! But smart self-consumption is even better, isn't it?

Calculate the profitability of your solar installation

It is necessary to study the degree of profitability of your project in its entirety. For this, I recommend the WattNext simulator. It automatically calculates the amount of your self-consumption premium, your profitability, and your earnings over 25 years.

And if you want to do the exercise yourself, here is the formula:

(Investment - Aid) + (Income generated over 25 years - Expenses to be expected) = The profitability of your project

We will apply it a little further down with concrete examples, but before that, it is necessary to understand each component of this calculation.

8(b). Investment: how much does it cost?

As an indication of the average prices observed on the market for a photovoltaic installation in superimposition in 2020, it is necessary to count:

Between $9,000 and $13,000 For an installation less than or equal to 3 kW

Between $14,000 and $18,000 For a 6 kW installation

Between $18,000 and $22,000 For a 9 kW installation

Attention: if the price offered to you is higher, run away! It is a scam.

The main parameters that determine the price:
1.The number of panels installed (i.e., the total power of the installation)

The type of inverter

The type of installation (labor, equipment, wiring, etc.)

Part of the costs of this photovoltaic installation is fixed, that is to say, that it does not increase with the number of panels installed. For example, the movement of installers, the time required to erect the scaffolding, the cost of the inverter and electrical connections, the time spent on administrative procedures, etc.

The more panels, the more they produce, and the more they accelerate the amortization of these fixed costs. This is why we should not be content to look only at the cost of the installation but to study its profitability in the long term. Moreover, when I ask the opinion of experts, their response is final:

The winning strategy is to put a maximum of panels within the limit of the space available because this reduces the cost price of the kWh produced.

2. Financial aid for solar self-consumption in 2020: what are you entitled to?

To facilitate the energy transition and accelerate the movement of renewables, the State now grants a premium for self-consumption, also called an investment bonus. This reduces the purchase price of your photovoltaic solar panels and its amount changes depending on the size of your installation. Its payment is annual and divided into five equal parts (1/5 of the amount will be paid each year for five years).

Since the last tariff decree in force in the 1st Quarter, you can claim to:

Premium amount Installation power

$390/KW Less than 3 KW

$290/KW Between 3 kW and 9 KW

$180/KW Between 9 kW and 36 KW

$90/KW Between 36 kW and 100 KW

Example: the Campbell family had 27 panels of 350 Wp installed, or (27 x 350 = 8750) an installation with a total power of 8750 Wp or 8,750 KW. Their premium will be $290/KW, i.e. $290 x 8,750 KW = $ 2,537.50.

The Campbells will receive $ 2,537.50 in 5 years (i.e. $ 507.50 / year).

Since 2014, exit the tax credit for photovoltaics replaced by Mine. Why? Because the premium for self-consumption and the purchase of the surplus of your electricity by EDF are sufficient incentives. Indeed, the sale of your surplus electricity is not negligible: there is a faster amortization effect.

Finally, there are also local aids to which you may be eligible. To find out if the municipality is concerned, contact the town hall, the place of installation, or your nearest ADEME energy center.

3. Income generated over 25 years

Once your panels have been installed, the connection request and the start-up have been completed, and your solar installation produces electricity. It allows you to amortize the cost of its installation with:

Savings on your electricity bills: all the electricity supplied by the panels and consumed in your house translates into an equivalent reduction on the amount of your electricity bill: 30 to 50% less, depending on the installed power and your consumption profile. Plus that's without even having a battery!

Example: the Campbell family is made up of four people and consumed 11,770 kWh annually, or $1,829.17 of electricity bill before installing solar panels. Since they directly self- consume the electricity supplied by their 25 panels, it is $ 1,079.21 (a saving of more than 41% on their bill, or $ 749.96/year).

Revenues from the sale of surplus: EDF OA (EDF Obligation de Achat) should buy your surplus electricity production at a fixed price guaranteed over 20 years, even if it is not the total sale of your production. All electricity that could not be consumed is sold. This generates an annual income, which will offset the residual invoice due to your supplier.

Purchase prices by EDF Obligation to Purchase in the 1st quarter of 2020:

Purchase price $/kWh Installation power

$ 0.10/kWh Less than or equal to 3 kW

$ 0.10/kWh Less than or equal to 9 kW

$ 0.06/kWh Less than or equal to 36 kW

$ 0.06/kWh Less than or equal to 100 kW

Example: the Campbell family sells their surplus production. This year, 6010 kWh will be sold for $ 0.10/ kWh. The Campbells, therefore, made a profit of $601 ($ 6,010 x $ 0.10 = $601) by selling its surplus production.

Please note: Note that above 9 KW, there is a loss of $ 0.04 per kWh sold. A loss which no doubt seems minimal, but which represents a shortfall of 40%!

Once the panels have been installed, maintenance is limited. The monitoring of the correct functioning of the panels is done by a monitoring application. In the event of a defect, you are automatically alerted by SMS or e-mail about the origin of the latter, and the diagnosis will be made easier by the after-sales service.

Finally, like your roof, the panels can receive pollution or dust, but the precipitation is generally sufficient to clean the dirt if the inclination is at least 15 °, and there is no specific pollution near your installation.

But there are still costs to be expected: the cost of access to the network invoiced by the supplier (the Turpe), your insurance if your housing contract does not support your installation and the possible replacement of your inverter within 25 years to come (an only electronic component of the installation).

Example: on our advice, the Campbells reserves $130 per year to cover these costs.

Be Transparent

The costs to be expected are not always explicitly mentioned in the offers of our colleagues. When you carry out a Solar Earth study, they are not only mentioned but also provisioned in a depreciation table to avoid unpleasant surprises. Because that's also what it is to be an activist.

4. Profitability

Now that you understand the composition of the costs and the gains that your photovoltaic installation gives you let's discover the cogs of profitability.

You now know that:

The more signs you have, the more you save.

Conversely, if your installation exceeds a power of 9 KW, your electricity is purchased 40% cheaper.

So, to ensure the profitability of your project, we recommend using your surface area as long as you can, taking care not to exceed 9 KW. This is to allow you to earn as much as possible while investing less.

Example: the Campbell family has an installation whose power is less than 9 KW. This year, it has therefore won:

$ 749.96 savings in self-consumption of own electricity (41% of their bill)

$ 601 in revenue from the sale of excess electricity (6,010 kW)

$ 507.50, i.e., the 1/5 annual premium for self-consumption

-130 $ of provisioned costs (Turpe, Insurance, Replacement of an inverter)

The total amount of the Campbells' earnings this year is $ 1,728.46.

And remember: their annual bill was $ 1,829.17 before self-consumption. Now their residual electricity expenditure is $ 100.71. ($ 1,829.17 - $ 1,728.46 = $ 100.71).

So, even if your system is not funded cash, the purchasing power thus released allows you to write this one in a few years.

When the installation is amortized? You take full advantage of its profitability: savings and income are always there from dawn until dusk.

8(c). Gain autonomy

We often imagine photovoltaic self-consumption as total energy autonomy: we would then become our own and sole supplier of electricity. This is partly true since you become your energy producer.

Towards total autonomy?

"Goodbye bills and nuclear energy!"

I would like to tell you that it is possible, but if the theory is simple, the practice remains utopian in 2020: it is necessary to reach an instantaneous balance between production and energy needs. This, therefore, requires drastically reviewing your habits of electrical consumption and equipping yourself with storage: this remains too expensive today to be profitable, but the technological progress underway in the field of hydrogen storage will make autonomy accessible to everyone in a few years.

Wait, rest assured: there are other ways to optimize its consumption, such as the Lithium-ion battery or even home automation solutions, which bring their lots of technologies to allow you to achieve this.

Note: separating yourself from the general network will deprive you of the possibility of selling the surplus of your production. This solution is only recommended in isolated areas where consumption is completely consumed; otherwise, it will be more prudent to leave your installation connected to the network.

Focus on solar batteries

The battery function is simple. It is used to store excess electricity at the peak of production, to use it later when your panels can no longer cover all of your needs. Some also use it with an emergency system so that in the event of a power outage, electrical devices remain operational, as do alarm systems.

Lithium-ion (Li-Ion) technology is the most efficient today. Under the impulse of the automobile manufacturers, it will continue to progress and allows today to store better than yesterday and more sustainably. But its cost represents an additional investment, and that is why storage systems are not always considered when installing solar panels. However, your installation can be designed to add a battery in the future.

For the safety of installations and the quality of storage over time, good quality batteries made in Germany or Austria with Sony or LG cells from $ 3,588, including tax, are recommended. These batteries are guaranteed for a charge cycle amount equivalent to around twenty years when used in conjunction with a photovoltaic system.

As you will no doubt understand, the solar storage revolution is underway!

What to conclude from it?

Photovoltaic self-consumption makes it possible to produce and consume renewable energy directly at home.

But above all, it allows us to invest in an economically and ecologically sustainable future by reducing polluting energies by creating our energy resources. Either, the most efficient way to work for the energy transition!

For you, for future generations and our planet. What are you waiting for to act?

9.

IS IT POSSIBLE TO HAVE CONNECTED ACCOMMODATION WITHOUT AN INTERNET CONNECTION?

Yes, having accommodation connected without an internet connection is a reality! But to benefit from all the advantages of connected objects, an ADSL connection is very useful. I have several clients who do not have ADSL. Most often, it is for their second home, because there is no telephone line, for economic reasons. I installed several routers with a 3G/4G key when the coverage is good; otherwise, satellite antennas or even simple GSM keys depending on the coverage. Without any connection to the outside world, the installation of a communicating system loses much of its meaning. For really basic heating type installations, a Simple socket may be sufficient to meet the need. It is a socket block that directly receives a Sim card, and which switches the device which is connected to it on or off according to the SMS received.

How does it work?

The home automation center sends and receives SMS via its Sim card to accredited telephones.

Why deprive yourself of the internet and use a GSM solution?

The first reason is cost. An ADSL box has a monthly subscription, which is not negligible when it is combined with a telephone line. For an alarm or a heating control, it is a budget that is not related. The second reason is technical: no possibility of connecting to a telephone line. The third reason is often not admitted. Customers who request a solution by SMS do not have a smartphone or are afraid of using it. An 80-year-old customer once asked me for an alarm with remote controls to activate it. He refused to use "this modern thing that my children gave me." After the installation, he saw me testing the alarm with my laptop. He shyly asked me if his smartphone could do that too. An hour later, and with long explanations, he put the remote controls in a drawer so that they would not come out again. "It's so much more practical."

For this generation, touching the digital world is a shock. I often encountered the same reluctance with younger people. You have to know how to be convincing and above all, stay present to answer questions, to explain again. I never touch my clients' cell phones. They do the manipulations themselves on their phone. They learn by doing. They discover that they can understand and use a technology that they have not seen before. It requires availability and sometimes patience, but I like this aspect of the job a lot.

Which telephone operator do you recommend?

To have accommodation connected without an internet connection, I recommend the telephone operator, which best covers the desired sector and the cheapest. A small note for frequent travelers: SMS in certain countries pass better than push notifications or emails.

What autonomous devices, without the need for an internet box do you recommend?

Far from having tested them all, I use the Jeedom Pro with the Zwave protocol and a Huawei key.

What functions can I take advantage of with devices operating on GSM?

An SMS triggers both a simple action and a complex scenario. The control unit can send an alert, states, and measured values. We must, of course, forget the consultation of graphics over a week, photos, videos. These data are reserved for ADSL lines.

Are there GSM-compatible home automation boxes, and what is the point of having one?

Yes, there are several. The Jeedom Pro is a wonderful toolbox. No installation and it uses all of the resources it offers. She has the answer to almost everything.

What should I add to my box so that it works on GSM?

A USB key containing an active Sim card.

What can I activate or deactivate with a home automation box by GSM?

You can send a temperature setpoint for heating, receive an intrusion, flood, low-temperature alert…what do I know? Each client has different objectives and specific requirements, depending on the project. For some, it is intrusion monitoring, for others, energy savings or comfort.

Is it possible to retrieve information from my connected accommodation without an internet connection, by SMS? If so, why?

Just send a preprogrammed question to get an answer. "What is the temperature of the living room?" "The living room temperature is 21.3 °". Smartphones write and read text messages. You don't even have to use the keyboard anymore. Home automation has experienced several revolutions. Since the electrical wiring and flocks of switches from the 60s and 80s, the house has successively switched to the KNX TP, which has become the latest trend for building automation.

Home automation revolutions followed with the internet first and above all, the arrival of smartphones, which made it possible to have "your home in your pocket." Now we are entering the era of voice control with connected speakers. The next revolution is already looming with artificial intelligence, which anticipates our needs before we have expressed them. The connected house is evolving very quickly. It was time! If we observe the majority of homes today compared to those of 30 years ago, apart from a room thermostat, electric shutters, and gate, an intercom, nothing has changed. Our cars in the same period were covered with electronics. Having a car key with a button to lock the doors is in itself pointless. We are all capable of manually locking the doors. Yet, no one today would do without this comfort. Home automation is the same thing. We are all able to close the shutters or turn off the radiator. Do we still have to think about it at the right time and be able to do it? As I speak, my shutters on the south facade have just lowered automatically because the sun is coming into the room. I didn't think about it anymore. The house will stay cool throughout the summer, without a noisy and energy-consuming air conditioner. Home automation allows you to focus your energy on more important things in life.

Can multiple users manage my connected accommodation without an internet connection, by SMS?

Yes, sharing is a big principle. For accommodation connected without an internet connection, it is important to specify that users can have different rights. In an Airbnb rental, the visitor does not have the same rights as the administrator.

How do I know if an order has been executed by SMS?

Often, I program two responses: the first to say that the SMS has been received by the central, the second to indicate that the order has been executed, with a measured value if necessary.

Similarly, if I sent the wrong SMS, will I have a return from my box?

Yes, in Jeedom, the control unit responds that the command is not understood.

If tomorrow I finally want to put an internet connection in my accommodation, could this box be useful for me?

It is the same. It will be enough to connect the RJ45 port of the Jeedom Pro to the RJ45 port of the ADSL box and activate the DNS for a remote link. A little true story: A Parisian client has equipped his second home in Annecy in Jeedom. Its Livebox often desynchronizes, thus cutting the remote link with the Jeedom. Orange advises her to disconnect and reconnect her Livebox so that she can resynchronize. From Paris, this is not possible. The Jeedom is then reachable via the Internet. So, I added a GSM key in addition to the link ADSL, and a controlled outlet on the power supply of the Livebox. When the Livebox desynchronizes, the client sends an SMS to the Jeedom, which restarts the Livebox, which resynchronizes. The Jeedom has connected again. It can, therefore, be very useful to have a GSM link in addition to the ADSL link.

10.

FINANCING YOUR ECOLOGICAL
HOME PROJECT

The reality of the market to obtain financing

With his company Belvedair, Benoît Lavigueur has built over a hundred ecological houses over the past ten years. He knows the realities of the market and his rich experience of entrepreneur and trainer, shared in his presentation to the Certificate in Design of Ecological Building of ERA Solution, allows to prepare strategically for obtaining financing to carry out a project of ecological construction.

Since the financial crisis of 2008, obtaining loans for the construction of ecological housing has become more difficult, because building a quality house, efficient and eco-affordable, costs one more at the moment - even if it is of a smart investment, while in the eyes of the lending bank, it has no recognized increased value. At least, not yet.

However, Benoît specifies that things are progressing little by little to change mentalities in financial institutions. Indeed, a Certificate student, who is also a certified assessor, understood the importance of changing the ways of doing things in this area. In her profession, she now takes into account the added value of an ecological house. This is great news that offers hope for the future!

Also, to meet the however surmountable challenge, insofar as we organize ourselves strategically, Benoît first insists on two essential points.

The purchase and full payment of the land

Before applying for a grant, it is crucial to have not only bought the land but to have paid for it entirely. Sometimes patience works for you, even if it takes a while to fully repay. As Benoît points out, 10 years ago, when you went to the bank for a loan and paid for your land, you easily received the money to build yourself, which is no longer the case today. What is more needed these days?

Ensure a liquidity amount between $ 30,000 and $ 50,000. Most construction projects that have failed along the way will be killed in the bud by a lack of liquidity and fluid budget during the construction phase. Besides, the presence of liquidity will help you gain the trust of the lending financial institution and on which your entire project will depend on its main funding.

The importance of liquidity

It is even fundamental! For example, it is much better to have money in your bank account that initiates a construction project, than to already own the household appliances that will be there, because the bank will not consider these goods as a cash value.

Even if you are very thrifty and are good managers of your

personal and family budget, the fact remains that the construction of your house will be the moment when you will need the most cash in your life. Any expense that can be carried over, carry them over! Don't change cars before you build! Every dollar in your account will make your building experience smoother.

However, for the average person, finding liquidity of $ 30,000 to $ 50,000 seems an impossible challenge. Do not panic! Benoît is the master of solutions.

Solutions offered

What Benoît has most at heart is that ecological projects succeed, to move towards a global ecological transition. So, he shares some tips. How to find this liquidity?

Of savings: a renovation project or construction must become the number one priority in your life if that's what you choose to do. If you do not have a staggering salary, saving each dollar may be a sacrifice on your lifestyle (travel, high standard of living, additional car, etc.), but it is for the reason that you are important to you.

A loan from relatives (family, friends, or others) notarized or not, with interest or not, depending on the agreement, may be offered. Paying interest to loved ones you love, rather than an anonymous financial institution, can lower the cost of sacrifice!

Knowing how to inspire those around you with a quality ecological project can assist you (in the form of a loan or donation), while also inspiring other people in this direction. For example, when Benoît and his wife built their ecological house in their mid-twenties, they were not wealthy. They were lucky to be able to borrow $ 20,000 from their parents for their project.

With a stable job, it was not that they lacked income, but they needed the liquidity necessary to carry out such a project. Once their house was built, they were able to repay these amounts without problems with their respective parents.

Benoît also offers an important tip that allows you to play on the gray areas when applying for a construction loan to a financial institution:

If the loan from your loved ones has not been registered as a notarial deed, you are completely legal not to declare it to the financial institution. If the bank sees only paid land and an absence of debts: this can only be good for your file. However, if the loan is formalized, the financial institution will be informed.

Do you have paid land, liquidity of $ 30,000 to $ 50,000, and a complete file with plans for your future building? The art of properly presenting your file to the lending institution will determine the future of your project. How to do it?

The credibility of the project and the file for obtaining a loan

A well-assembled file will make all the difference, according to Benoît's experience. A financial manager will be sensitive to the terms used to describe your project, which could even make it difficult to obtain the loan. When planning a more alternative project, which can often be the case for the construction of ecological houses, it is important to be vigilant about what is presented. The banking environment is, by definition, a conventional environment based on solid assets and established and profitable traditions.

There is no point in worrying them with techniques that will still be unknown to them, such as straw or hemp insulation, even if they are very effective techniques! The assessor, as we will see below, just needs to know that the building will be isolated, period! In any case, construction techniques are not their responsibility. They are only there to assess how much the building will be worth, how much they will lend you (80% of this assessment) and establish the terms of payment.

At Belvedere, Benoît was able to observe several clients whose projects failed during the presentation to the finance manager. The project was more than viable; they just mentioned too many details that worked against them. Instead, speak of a "high-performance house:" a house that will cost you less in heating, which, in their eyes, will increase your repayment capacity! There you get their attention.

The challenge of formulation and presentation is not only played in presenting the project to a leading financial institution, but also to other official officials that you will have to convince of the viability of your project, whether the banker, municipal assessor, contractor, sub-contractors, friends, family, etc.

The trick is to always ask yourself what this person wants to hear, how to reassure them by being completely honest. Just like a salesperson, whose approach varies from one customer to another, know how to put yourself in the shoes of the person facing you.

Mortgage pre-approval

Depending on your income, the financial institution will grant you pre-approval for the project. This is usually a large amount. But beware! The amount of a pre-approval is not the amount of the loan you will get. This nuance is often misunderstood, which can have catastrophic consequences for the realization of the project.

For example, if the bank pre-approves you for $ 350,000, that doesn't mean it's the amount it will loan you. Rather, it is the amount you can survive daily - and the word is right - with a mortgage of $ 350,000. Shade!

Evaluation and obtaining funding

Let's take a concrete example to illustrate the process. You show up at the home financing company (HFC) with sufficient and stable income, land paid, and no debt - an official at least - cash of $ 30,000 to $ 50,000 and a house plan to build. You then get a pre-approval of $ 350,000.

If on the plans for your well-insulated, eco - affordable, high-performance ecological house, the construction costs amount to $ 300,000, the HFC will not offer you a loan of this value. Rather, it will assess the value of the house on the market.

It must be understood that the latter does not want to finance $ 300,000 for a house that would be worth $ 250,000 on the market. Because if you are no longer able to pay for it, they will not be able to recover this shortfall of $50,000.

Let's imagine, in our example, that the HFC values your project at $ 217,000. This market assessment covers the whole house and the land as a whole.

You know full well that the real cost of building the ecological house and that of the land is much higher than the amount assessed by the HFC because the latter always assesses the house at 80% of its real cost, to protect themselves and leave themselves room for maneuver.

The HFC, therefore, accepts to lend you 80% of this amount of $ 217,000. That is to say, $ 173,000, to follow our example. The budget is tightening more and more, even if we are in self-construction and that we receive a little help beside.

Loan payments in successive stages

But that's not all. The construction loan amount of $ 173,000 will be divided by construction stages. Indeed, only after the creation of a path on the ground leading to the dwelling and following the construction of the foundation, 22% of the amount can be granted to you. Once the frame is assembled, another amount "x" will be granted to you, etc. The grids are generally fairly standard, and each stage of construction is standardized.

The problem with this way of doing things is that the steps do not follow the reality of the costs generated for the auto-builder. For example, a foundation can be worth, say 11%, and painting say 5%. A self-builder will quickly realize that painting a home certainly does not cost half the cost of a foundation! However, once the painting is done in a building, it takes on much more value in the event of foreclosure and resale.

The investment for an efficient building will not be taken into account

Another fundamental element to grasp is that the HFC appraiser will only have three hours to produce a market assessment of your house and land project. Within these three hours, the latter must:

Study the plan

Complete an evaluation grid

Find three comparisons of equivalent houses in the area

It is understood that this one does not have time to go into the details of the plan of the house and the project!

Its evaluation is summarized on a single sheet of paper on which will be specified: the exact measurements of the habitat and whether the habitat is isolated or not! No nuance can be brought to this last point - that it is a high level of insulation, a low annual heating cost thanks to a high-performance house in energy terms, etc.

The fact that housing costs $ 200 instead of $ 2,000 of annual heating are not a factor in the value of the house. The added value of the high cost of insulating windows or ultra-efficient insulating materials is not taken into account. The thousands of dollars more that you invest in quality insulation will be overlooked in its evaluation!

15% withholding until the end of the project in the event of legal mortgages

Another essential detail: the 15% rule. With each promised payment, the HFC will retain 15% of the amount loaned to you. This 15% is retained until 35 days after the end of the housing construction project. The reason for this withholding is that the HFC is thus protecting itself from "legal mortgages." What does it consist of?

It is a right of subcontractors to protect themselves against bad payers. For example, if you have not paid the electrician who came to install your wiring, the latter has the right to file a "legal mortgage," which will be registered and notarized. The HFC cannot grant you the full amount of its loan until this legal mortgage has been paid. Why?

Legal mortgages have priority. If ever the house were to be liquidated, the subcontractor would be paid before the HFC, which is not to the advantage of the lending institution. So, it simply protects itself from legal mortgages. How?

All the subcontractors of your project have legally 30 days after the end of your work to deposit a legal mortgage if they have not been paid. The delay of 35 days after the end of the construction of the house, to pay you the full amount due, is a way for the HFC to fully protect itself from this type of situation and to only pay the amount due once the risk of 'legal mortgages passed.

Liquidity to complete construction

If we go back to our example, the HFC which had agreed to lend you $ 173,000, in fact only lends $ 150,000 during construction itself, because of $ 23,000, which corresponds to the 15% retained, will only be paid out 35 days after completion of construction of the complete building.

You see the picture from here, the reality of the self-builder is that it often happens that at this stage, cash funds become urgent and it is during this period that savings or loans from families or friends from $ 30,000 to $ 50,000 become invaluable resources during the construction process, up to 35 days after the end of it. Once the precious 15% is finally paid to you by the HFC, you can finally breathe and reimburse your loved ones.

In short, Benoît offers a lucid look at the reality of building a new ecological house in Quebec, which is more difficult than in the dream of many self-builders. However, as he would like to emphasize, once they are well aware of the realities of the market and the steps to follow, a solid project, well planned and well presented, can be carried out without problems!

10(a). Ecological house: state aid

To encourage the population to embark on the construction of ecological houses, different states grant various aids to reduce the fairly high investment cost.

Green/off-grid house: tax credits

When building your ecological house, you benefit from 5.5% VAT and a tax credit if you hire a professional.

The equipment concerned by the tax credit is low-temperature boilers 15% credit, condensing 25 to 40%, heat insulation of pipes, and thermal insulation materials 25 to 40%. It is also possible to benefit from a tax credit of 25 to 40% for heating control and programming devices. These are used to limit consumption without affecting comfort.

Renewable energy is also affected by a 40% tax credit: heaters, heat pumps, solar water heaters, wood heaters, or the heaters stove if the performance is sufficient. Photovoltaic solar panels, wind turbines, and hydraulic installations benefit from a tax credit of 50%. However, the amount of expenses giving entitlement to this tax credit is capped at $ 8,000 for a single individual and $ 16,000 for a couple over a maximum period of five years.

Ecological/off-grid house: the bonuses

When building an ecological house, the National Housing Agency grants bonuses. To benefit from it, the equipment used must be efficient. The aid varies: the premiums can amount to $ 900 for a heat pump, $ 900 for a condensing boiler, or even $ 80 for a window.

Homeowners can also benefit from an ecological loan at zero rates if they meet two of the following conditions: insulation of openings (window and doors), thermal insulation of exterior walls, insulation of the roof, production of hot water via energies or a heating installation operating on renewable energy. The loan is spread over ten years and is capped at $ 30,000.

10(b). Building an off-grid/ecological house: financial aid

44%. This is what the building sector represents in terms of energy consumption in France. Faced with the rise in energy prices (heating or electricity) and the growing awareness of environmental challenges, ecological construction is becoming a real necessity.

Green/off-grid building has many advantages: better energy performance, housing designed for the health and well-being of its inhabitants, or construction that respects the environment. However, many still raise the flagship problem of ecological constructions: their costs. Indeed, ecological construction is more expensive than conventional construction (10 to 20% more expensive, depending on the solutions chosen).

If this higher cost is offset over time by lowering charges for the house, there are other financing solutions to embark on the ecological adventure by reducing the cost of construction. What are the tax advantages for the construction of a new house, the aid for an ecological house? Find out about the solutions offered today by the State, local authorities, and other organizations.

Financial assistance for the construction of an ecological house-1

State aid

The first actor allowing you to make savings for the construction of your ecological house is the State. Faced with the need to build more intelligently, different states/governments have set up financing solutions to help individuals and professionals planning to build a house that respects the environment.

The Zero Rate Loan (ZRL)

It is quite simply the possibility of borrowing a certain sum of money without having to pay interest (loan at rate 0).

This aid is not directly intended for ecological constructions; however, the zero-rate loan can be very advantageous for first-time buyers.

The ZRL is granted to borrowers regardless of their level of resources if:

- They plan to build a house or buy a new property (never inhabited).

- They buy an old property and wish to carry out renovation work.

- They want to transform a room (which they already own) into housing.

In any event, the real estate concerned must be the principal residence of the borrower.

Regarding the amount of this loan and its duration, it will depend on your income, the location of the accommodation, or the number of occupants. To find out more, visit the government's ZRL website.

Exemption from property tax

Choosing a new house means being able to benefit from an exemption from property tax for two years! This exemption is very interesting and should convince buyers to opt for new housing and, therefore, more environmentally friendly. The exemption may even last longer for houses complying with BBC standards and the 2012 thermal regulations.

CITE: aid for the renovation

Do you already own a house and want to make it more ecological? It's possible. Many states, e.g., the French State indeed offer a tax credit (CITE: Tax Credit Energy Transition) to all owners wishing to carry out ecological renovation work for a house over two years old.

Financial assistance for the construction of an ecological house-2

Aid granted by local authorities

In addition to financial aid granted by the State, it is also possible to benefit from subsidies from local authorities. It can, for example, be aid from the Regional Council or even directly from the municipality. For example, we can cite the Rhône Alpes region, which benefits from a "Wood Plan" and a "Solar Plan." Thus, the region offers subsidies to individuals investing in wood, solar thermal energy, or solar photovoltaic energy.

To be able to benefit from these possible aids, we advise you to inquire at the town hall of your commune.

Other helpers to help you build an ecological house

In addition to aid granted by the State and local authorities, individuals also have the possibility of receiving financing solutions through organizations or associations working in favor of ecological transition.

The buildability bonus

First of all, it is possible to benefit from a constructability bonus of 30% maximum when building an ecological house.

Indeed, the body issuing the building permit can grant bonuses for housing:

Very energy efficient,

Low environmental impact,

Positive energy and

Bonuses granted by associations.

Finally, certain associations can grant bonuses to the purchase of renewable energy equipment (heat pump or still condensing boiler) for energy improvement works. Other organizations have a budget to help the ecological transition.

11.

THE DIFFERENT ECOLOGICAL/OFF-GRID HOUSES

In the construction of new houses, there are now several types of ecological houses. Before making your choice of builder, here are the different ecological houses and their usefulness for the environment. Overview of the different green/off-grid houses.

BBC house: the most widespread

Today, 17% of new homes built in France meet the BBC standard, Low Consumption Building. This BBC label aims to set a maximum limit on the energy consumption of new buildings for five primary uses: heating, lighting, ventilation, hot water, and air conditioning. Depending on the altitude, the geographical area, the living area, the new BBC house will have to consume, on average, 50 kWh of primary energy per m² and annually. From January 1, 2013, all residential building permits must correspond to this standard; this will be the new thermal regulation for all buildings, RT 2012.

Bioclimatic house: using the resources of the place

The bioclimatic house uses to its advantage the climate and the environment of the place of its establishment. The bioclimatic house could be compared to simple common sense. Examples: orient the new house in the south to heat it for free, use deciduous trees to provide shade in the summer, install the living rooms in the south, when the rooms in the north serve as an energy buffer.

As an alternative to heating and air conditioning, the bioclimatic house tries to leave aside heating systems deemed too expensive, such as solar collectors and other heat pumps. On the other hand, it favors building and insulation materials allowing temperature stabilization, both by the accumulation of energy from the sun in winter and by keeping it cool in summer. The bioclimatic design is notably used for the construction of a building with a high environmental quality (HQE) approach.

Passive or very low energy house

The principle of the passive house is that the heat released by the interior of the house (inhabitants and appliances) and that provided by the outside (sunshine) is enough to heat the house. Most passive houses are found in the countries of northern Europe, Germany, and Switzerland.

IAs for the bioclimatic house, insulation is the basic principle of the passive house, the credo of the passive house is to avoid thermal bridges, heat loss in away. Another important point of a passive house, managing to enhance the contribution of the sun, thanks in particular to the many glazed parts. Count between 40 and 60% of glass surfaces in the south! An additional construction cost of 20% must be provided in France to obtain a passive house.

Positive house: producing more energy than necessary

A positive house produces more energy than it consumes. The calculation of the energy consumption of a positive house is established over a long period, a year in general. These positive energy buildings are passive houses associated with energy production units, such as photovoltaic collectors on the roof, solar heating, a heat pump, or even a wood boiler. These positive new houses require significant financial investment at the time of construction.

The positive new house could correspond to the future standards of the thermal regulation of 2020.

11(a). How to secure an ecological house?

The ecological house is attracting more and more people. Indeed, this type of housing is healthy and offers good energy performance. For example, the low-consumption house, also known as the BBC, which emits a very low rate of green/off-grid house gases. As a reminder, the ecological house is built from materials that respect the environment. However, just like the classic house, it is important to secure it against burglary.

PVC burglar-resistant doors to secure the entrance to a home

It is important to reinforce the doors of an ecological house to benefit from secure housing. The installation of burglar-resistant PVC doors is ideal. From an eco-friendly perspective, the choice of this material is completely defensible. Indeed, its production requires only a small amount of water, salt, and a little oil, unlike other synthetic materials. PVC security doors considerably complicate the forcing of the door by thugs. In general, they always include one or more door cylinders that are difficult to force, as well as a secure door handle. Note that PVC can be recycled into tables or benches after serving for 50 years.

Install an alarm to avoid intrusions

To secure your ecological home, an alarm brings an undeniable plus. The role of this security system is to deter attempted break-ins and to alert the neighborhood. There are many kinds of alarms. Audible alarms alert people in the vicinity with a siren. They can also serve as an anti-intrusion system, combined with a system sending a signal to owners and the competent authorities. Connected alarm systems are more efficient than the first type. They use a connection and warn the owners of the house of situations that could require its intervention. This type of security system notifies on the smartphone what is happening in the home, and this, in real-time.

Monitor your ecological home permanently with a video surveillance system

CCTV is a system for monitoring your home remotely. It is equipped with cameras that capture the images. You have the choice between analog video surveillance and IP video surveillance. The first is the most classic device. IP video surveillance, for its part, is equipped with cameras linked to a computer network that processes images. In other words, the images are found on a secure internet network. To install the device, you will not need professionals. You will find a wide selection of video surveillance kits in the market. An ecological house with a video surveillance system will be completely secure. So, you can calmly go on vacation or work.

11(b). Sustainable housing: five tips for building a self-contained house

You have dreamed of it for years: to build a house that looks like you, by following your environmental values. In the end, this fantasy is not that complicated to realize. All the criteria and advice are that you need to build a sustainable habitat that respects nature.

1. Use natural insulation

Insulating a house, whether it is self-contained or not, is fundamental. During construction, this step is crucial. Insulation allows you to avoid humidity, cold, heat, and energy bills that can climb at high speed! For a house that is more respectful of the environment and well-insulated to be energy independent, your house can be imagined with bio-based insulating materials.

We distinguish among natural insulators:

- vegetable, animal and mineral fibers: rock wool, glass, wood, flax, hemp, and sheep;

- renewable materials: cork, recycled paper;

- mineral insulators: perlite and cellular glass.

These natural insulators allow optimal thermal and sound insulation.

If they are more difficult to find and more expensive than "conventional" insulation, they remain particularly effective and efficient, which is THE condition for making your home independent.

Alexandre Vasquez, the thermal engineer in the Synergisud thermal design office, insists on the future:

"The public authorities have taken into consideration the importance of the level of green/off-grid house gas emissions of a building over its entire life cycle (production of materials, their transportation, their implementation, their life and their treatment at the end of life). As such, for all new construction from 2021, an environmental assessment of these emissions over the entire life cycle of the building will have to be carried out by a competent design office, taking into account the carbon footprint of each of its constructive components and technical equipment envisaged."

To know

The autonomous house is a so-called alternative habitat that depends on specific legislation that you can discover on the website of the Ministry of Territorial Cohesion.

2. Install photovoltaic panels

The installation of solar cells or photovoltaic panels is a fundamental step in the construction of a self-contained house. Solar cells (installed outside your home), transform light energy into electricity, making your house self-sufficient in electricity.

A solar cell is made up of several small photovoltaic panels that collect solar radiation and store it thanks to a battery that converts it into volts. This system, therefore, makes it possible to produce electrical energy from solar radiation. They become low-voltage direct current, around 12 to 24 volts, then can be transformed by a converter into a domestic current of 220 or 230 volts.

These photovoltaic panels can be supplemented by a wind turbine, which provides additional energy, especially in winter, when the sun is less present. The lifespan of solar cells is estimated to be around 20 years, at a price ranging from 2,000 to 3,000 euros per square meter installed. A reasonable investment for significant freedom.

3.Collect rainwater

Using rainwater is part of the crusade to live in a self-contained house. Both economic and ecological, the recovery of rainwater allows you not to depend entirely on the public network and thus gain independence.

But what does the law say about it? The decree of August 21, 2008, published in the Official Journal n° 0201 of August 29, 2008, legislates the recovery and use of rainwater because "they can contain pathogenic micro-organisms."

1. To be able to harvest the rain and use it:

2. You should use filters, even if you only use it to clean your exteriors.

3. You can first have a filter at the gutter of your house. In the form of a grid, you avoid the maximum of external debris. At this point, water can be used for your exteriors.

4. A primary filtration can then be installed just after the pump, to use the water for the toilets or the cleaning of the floors inside the house. Just connect your water supply to the house piping.

5. Secondary filtration by adding an activated carbon filter after the water heater will be necessary to use the water for body washing, for example. For the purification of rainwater, it will be necessary to add a final step with a UV sterilizer or a ceramic cartridge, for example. For these two steps, call a

professional (a plumber, for example).

4. Install a dry toilet

Use drinking water for your toilet? An aberration for the followers of "dry" toilets, who therefore do not use water. 20 % of the water used by a household is dedicated to sanitary facilities. "More and more individuals are installing dry toilets in their homes! A less polluting solution due to the reduction in the volume of wastewater," says Alexandre Vasquez.

As economical as it is ecological, dry toilets operate on the principle of composting. Installed outside or inside the house, they still suffer from many prejudices. However, "new generation" dry toilets are gaining reliability and appealing to many ecologists!

Concretely, it is a seat below which a bucket is installed. Comfort is entirely compatible with this system: it all depends on the materials you want to use. The bucket is filled with sawdust or wood shavings. What is collected is absorbed by these materials, which cancel out odor and fermentation.

Make sure to empty it at least every week, either in a dedicated place in your garden or in a compost that you have made, based on soil, which will also serve as a kitchen trash can. This is called the material composting system.

Good to know

It is only since 2009 that dry toilets have been authorized in France. The decree noted that "they should not generate any nuisance to neighbors or liquid waste outside the plot or pollution of surface water and groundwater."

5. Use a solar water heater

These are solar panels that heat your water! Their smooth surface receives the rays of the sun, which heat a fluid present under the panels. The hot fluid is stored in a tank that is connected to the sanitary facilities, taps, sinks, and showers in your home.

Before any installation, be sure to check the angle of your panel relative to the sun, regardless of its installation: in your garden or on your roof. You can call on a professional to benefit from an impeccable pose.

To go even further in the desire for energy autonomy, the thermal engineer Alexandre Vasquez even adds, "Why not also recover the heat from the hot water in your shower before it goes down the drain?" In recent years, technologies have been used to capture the heat from wastewater, and this heat is naturally transformed into energy to heat your next shower.

CONCLUSION

Off-grid housing may be the dream of many and is slowly gaining popularity among people concerned about environment-friendly habitat. However, gaining total autonomy should be a gradual and stepwise process, starting with energy conservation. Many factors play an important role in this regard, from the choice of energy-saving appliances to different materials chosen for the construction of the house itself.

Off-grid/autonomous hose construction may cost much more than a conventional home construction. However, it proves a considerable saving in the long run. Think and plan carefully, before making the conversion, seeking professional help at every stage is highly recommended, to avoid problems later.